Studies in Computational Intelligence

Volume 643

Series editor

Janusz Kacprzyk, Polish Academy of Sciences, Warsaw, Poland
e-mail: kacprzyk@ibspan.waw.pl

About this Series

The series "Studies in Computational Intelligence" (SCI) publishes new developments and advances in the various areas of computational intelligence—quickly and with a high quality. The intent is to cover the theory, applications, and design methods of computational intelligence, as embedded in the fields of engineering, computer science, physics and life sciences, as well as the methodologies behind them. The series contains monographs, lecture notes and edited volumes in computational intelligence spanning the areas of neural networks, connectionist systems, genetic algorithms, evolutionary computation, artificial intelligence, cellular automata, self-organizing systems, soft computing, fuzzy systems, and hybrid intelligent systems. Of particular value to both the contributors and the readership are the short publication timeframe and the worldwide distribution, which enable both wide and rapid dissemination of research output.

More information about this series at http://www.springer.com/series/7092

Kun Ma · Ajith Abraham
Bo Yang · Runyuan Sun

Intelligent Web Data Management: Software Architectures and Emerging Technologies

Kun Ma
Shandong Provincial Key Laboratory
 of Network Based Intelligent Computing,
 School of Information Science
 and Engineering
University of Jinan
Jinan, Shandong
China

Ajith Abraham
Scientific Network for Innovation
 and Research Excellence
Machine Intelligence Research Labs
 (MIR Labs)
Auburn, WA
USA

Bo Yang
Shandong Provincial Key Laboratory
 of Network Based Intelligent Computing,
 School of Information Science
 and Engineering
University of Jinan
Jinan, Shandong
China

Runyuan Sun
Shandong Provincial Key Laboratory
 of Network Based Intelligent Computing,
 School of Information Science
 and Engineering
University of Jinan
Jinan, Shandong
China

ISSN 1860-949X   ISSN 1860-9503 (electronic)
Studies in Computational Intelligence
ISBN 978-3-319-80745-4   ISBN 978-3-319-30192-1 (eBook)
DOI 10.1007/978-3-319-30192-1

Printed on acid-free paper

This Springer imprint is published by SpringerNature
The registered company is Springer International Publishing AG Switzerland

Preface

The goal of this book is to present the methods of intelligent Web data management, including novel software architectures and emerging technologies and then validate this architecture using experimental data and real-world applications. Furthermore, the extensibility mechanisms are discussed. This book is organized to blend in with the research findings of the author in the past few years.

The contents of this book are focused on four popular thematic categories of intelligent Web data management: cloud computing, social networking, monitoring and literature management. There are a number of applications in these areas, but there is a lack of mature software architecture. Having participated in more than 20 software projects in the past 10 years, we have some interesting experience to share with readers. Therefore, this book attempts to introduce some new intelligent Web data management methods, including software architectures and emerging technologies. The book is organized into four parts as detailed below.

Part I: Cloud Computing

Part I introduces intelligent Web data management in the area of cloud computing. This part emphasizes some software architectures of cloud computing.

Chapter 1 deals with intelligent Web data management of multi-tenant data middleware. This chapter introduces intelligent Web data management of a transparent data middleware to support multi-tenancy. This approach is transparent to the developers of cloud applications.

Chapter 2 presents intelligent Web data management of NoSQL data warehouse. This chapter introduces intelligent Web data management of NoSQL data warehouse, which is used to address the issue of formulating no redundant data warehouse with small amount of storage space for the purpose of their composition in a way that utilizes the MapReduce framework. The experiments are illustrated to successfully build the NoSQL data warehouse reducing data redundancy compared with document with timestamp and lifecycle tag solutions.

Part II: Social Networking

Part II of this book introduces intelligent Web data management in the area of social networking. This part emphasizes some software architectures for social networking.

Chapter 3 presents intelligent Web data management of social question answering. This chapter introduces intelligent Web data management of a question answering system, which aims at improving the success ratio of the question answering process with a multi-tenant architecture.

Chapter 4 deals with intelligent Web data management of content syndication and recommendation. This chapter introduces intelligent Web data management of a content syndication and recommendation system. The experimental result depicts that the developed architecture speeds up the search and synchronization process, and provides friendly user experience.

Part III: Monitoring

Part III of this book introduces intelligent Web data management in the area of monitoring. This part emphasizes some software architectures for intelligent monitoring.

Chapter 5 presents intelligent Web data management infrastructure and software monitoring. This chapter introduces intelligent Web data management of a light-weight module-centralized and aspect-oriented monitoring system. This framework performs end-to-end measurements at infrastructure and software in the cloud. It monitors the quality of service (QoS) parameters of the Infrastructure as a Service (IaaS) and Software as a Service (SaaS) layer in the form of plug-in bundles. The experiments provide insight into the modules of cloud monitoring. All the modules constitute the entire proposed framework to improve the performance in hybrid clouds.

Chapter 6 deals with intelligent Web data management of WebSocket-based real-time monitoring. This chapter introduces intelligent Web data management of a WebSocket-based real-time monitoring system for remote intelligent buildings. The monitoring experimental results show that the average latency time of the developed WebSocket monitoring is generally lower than polling, FlashSocket and Socket solution, and the storage experimental results show that our storage model has low redundancy rate, storage space and latency.

Part IV: Literature Management

Part IV of this book introduces intelligent Web data management in the area of literature management. This part emphasizes some software architectures of literature management.

Chapter 7 illustrates intelligent Web data management for literature validation. This chapter introduces intelligent Web data management of a literature validation system, which aims at validating the literature by the author name from the third-party integrated system and the metadata from the DOI content negotiation proxy. The analysis of application's effect shows the ability to verify the authenticity of the literature by the author name from the system and the metadata from our DOI content negotiation proxy.

Chapter 8 presents intelligent Web data management for literature sharing. This chapter introduces intelligent Web data management of a bookmarklet-triggered unified literature sharing system. This architecture allows easy manipulation of the literature sharing and academic exchange, which are used frequently and are very often necessary in scientific activity such as research, writing chapters and dissertations, and preparing reports.

This book is written primarily for academic researchers who are interested in intelligent Web data management of some emerging software systems, or software architects who are interested in developing intelligent software architecture in the aspect of Web data management. However, it was also written keeping in mind the postgraduates who are studying Web data management. We assume basic familiarity with the concepts of Web data management, but also provide pointers to sources of information to fill in the background.

Many people have collaborated to shape the technical contents of this book. Our thanks to our colleagues for the wonderful feedback, which helped us to enhance the quality of the manuscript. We also thank the Springer Series on Studies on Computational Intelligence Editorial Team: Prof. Dr. Janusz Kacprzyk, Dr. Thomas Ditzinger and Mr. Holger Schaepe for the wonderful support to publish this book very quickly.

We hope the readers will enjoy the contents and we await for further feedback to further improve the work.

Kun Ma
Ajith Abraham
Bo Yang
Runyuan Sun

Contents

Part III   Monitoring

Part I
Cloud Computing

Chapter 1
Intelligent Web Data Management of Multi-tenant Data Middleware

1.1 Introduction

1.1.1 Background

Software as a service (SaaS), is a software delivery model in which software is deployed as a hosted service and accessed over the Internet [1]. SaaS has become the words that are on everyone's lips, and is going to have a major impact on software industry. According to International Data Corporation's (IDC) latest market report, SaaS will grow at a 26.6 % annual compound rate through 2014–2018 [2].

Today, SaaS applications are expected to take advantage of the benefits of centralization through a single-instance as well as multi-tenant architecture, and to provide a feature-rich experience competitive with comparable applications. A typical SaaS application is offered either directly by the vendor or by a service provider. In contrast to the one-time licensing model commonly used for a software, SaaS application access is frequently sold using a subscription model, with customers paying an ongoing fee to use the application. Fee structures vary from application to application. For example, some providers charge a flat rate for unlimited access to some or all of the application's features, while others charge varying rates that are based on usage.

Multi-tenancy is the most significant paradigm of SaaS, which is different from traditional software [3]. Multi-tenancy refers to a principle in software architecture where a single instance of the software runs on a server, serving multiple tenants. It contrasts with multi-instance architectures where separate software instances operate on behalf of different client organizations. In cloud computing, the meaning of multi-tenant architecture has broadened because of new service models that take advantage of virtualization and remote access. A SaaS provider, for example, can run one instance of its application on one instance of a database and provide Web access to multiple tenants. In such a scenario, each tenant's data is isolated and remains invisible to other tenants.

© Springer International Publishing Switzerland 2016
K. Ma et al., *Intelligent Web Data Management: Software Architectures and Emerging Technologies*, Studies in Computational Intelligence 643,
DOI 10.1007/978-3-319-30192-1_1

Multi-tenant application is regarded as a potential segment and the utilization in the aspect of enterprise applications, such as enterprise resource planning, office automation, and e-business [4]. However, it is difficult to manage the Web data of multi-tenant applications. In this chapter, we introduce intelligent Web data management of a multi-tenant application using a transparent data middleware.

1.1.2 Challenges and Contributions

There are some challenges of current multi-tenant data techniques. First, how to make the data middleware transparent to the developers is more challenging. That is to say that the legacy application is assured to migrate to the multitenant one with minimum modification of the source codes. Second, how to minimize the cost and impact of the database performance is also challenging.

To address these challenges, we introduce the architecture of a transparent data middleware to support multi-tenancy. The architecture of this data middleware is discussed in detail. The contributions of this data middleware are several folds. First, the data middleware is transparent to the developers. It is easy to make the legacy application to support multi-tenancy without re-architecting the entire system from the ground up. Second, some auxiliary optimized measures of the architecture are added to make this data middleware more extensive and scalable.

1.2   Related Work and Emerging Techniques

This Section introduces the related work and techniques on multi-tenant data middleware.

1.2.1 Software as a Service Maturity Model

Broadly speaking, SaaS application maturity can be expressed using a model with four distinct levels [5]. Each level is distinguished from the previous one by the addition of scalability, multi-tenancy, and configuration. Figure 1.1 shows the SaaS maturity model.

Level 1: Ad Hoc/Custom
The first level of maturity is similar to the traditional application service provider (ASP) model of software delivery, dating back to the 1990s. At this level, each tenant has its own customized version of the hosted application, and runs its own instance of the application on the host's servers. Architecturally, software at this maturity level is very similar to traditionally-sold line-of-business software.

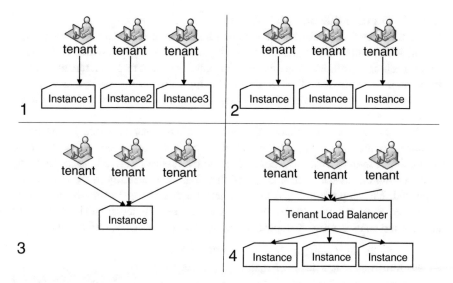

Fig. 1.1 SaaS maturity model

Typically, traditional client-server applications can be moved to a SaaS model at the first level of maturity, with relatively little development effort, and without re-architecting the entire system from the ground up. Although this level offers few of the benefits of a fully mature SaaS solution, it does allow vendors to reduce costs by consolidating server hardware and administration.

Level 2: Configurable

At the second level of maturity, the vendor hosts a separate instance of the application for each tenant. Whereas in the first level each instance is individually customized for the tenant, at this level, all instances use the same code implementation, and the vendor meets tenants' needs by providing detailed configuration options that allow the tenant to change how the application looks and behaves to its users. Despite being identical to one another at the code level, each instance remains wholly isolated from all the others.

Moving to a single code base for all of a vendor's tenants greatly reduces a SaaS application's service requirements, because any changes made to the code base can be easily provided to all of the vendor's tenants at once, thereby eliminating the need to upgrade or slipstream individual customized instances. However, repositioning a traditional application as SaaS at the second maturity level can require significantly more re-architecting than at the first level, if the application has been designed for individual customization rather than configuration metadata. Similarly to the first maturity level, the second level requires that the vendor provide sufficient hardware and storage to support a potentially large number of application instances running concurrently.

Level 3: Configurable and Multi-tenant
At the third level of maturity, the vendor runs a single instance that serves every tenant, with configurable metadata providing a unique user experience and feature set for each one. Authorization and security policies ensure that each tenant's data is kept separate from that of other customers logically. From the end user's perspective, there is no indication that the application instance is being shared among multiple tenants.

This mature model eliminates the need to provide server space for as many instances as the vendor has customers, allowing for much more efficient use of computing resources than the second level, which translates directly to lower costs. A significant disadvantage of this approach is that the scalability of the application is limited. Unless partitioning is used to manage database performance, the application can be scaled only by moving it to a more powerful server (scaling up), until diminishing returns make it impossible to add more power cost-effectively.

Level 4: Scalable, Configurable, and Multi-tenant
At the fourth and final level of maturity, the vendor hosts multiple tenants on a load-balanced farm of identical instances, with each tenant's data kept separate, and with configurable metadata providing a unique user experience and feature set for each tenant. A SaaS system is scalable to an arbitrarily large number of tenants, because the number of servers and instances on the back end can be increased or decreased as necessary to match demand, without requiring additional re-architecting of the application, and changes or fixes can be rolled out to thousands of tenants as easily as a single tenant.

1.2.2   Software as a Service Data Models

The distinction between shared data and isolated data is not binary. Instead, it is more of a continuum, with many variations that are possible between the two extremes. Therefore, there are mainly three SaaS data models from the balance between isolation and sharing [6–8]. Figure 1.2 shows the current SaaS data models.

Model A: Separate application and separate database
Separate application and separate database uses different separate applications and databases for each tenant, which is the simplest approach to data model. Unfortunately, this approach tends to lead to higher costs for maintaining equipment and backing up tenant data. The number of tenants that can be housed on a given database server is limited by the number of databases that the server can support.

Model B: Shared application and separate database
In this model, computing resources and application code are generally shared between all the tenants on a server, but each tenant has its own set of data that remains logically isolated from data that belongs to all other tenants. Metadata

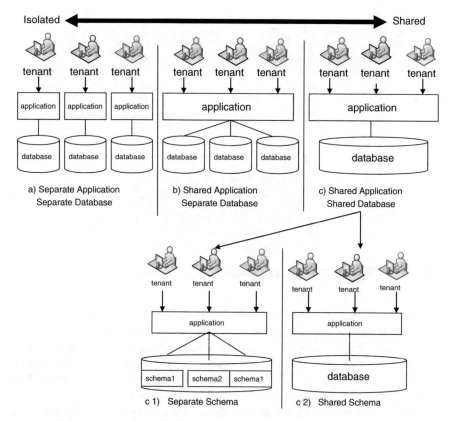

Fig. 1.2 Current SaaS solutions

associates each database with the correct tenant, and database security prevents any tenant from accidentally or maliciously accessing other tenants' data.

Giving each tenant its own database makes it easy to extend the application's data model to meet tenants' individual needs, and restoring a tenant's data from backups in the event of a failure is a relatively simple procedure. Unfortunately, this approach tends to lead to higher costs for maintaining equipment and backing up tenant data. Hardware costs are also higher than they are under alternative approaches, as the number of tenants that can be housed on a given database server is limited by the number of databases that the server can support.

Model C: Shared application and shared database

From the aspect of fine-grained partition of shared data model, there are two shared SaaS data models: separate schema and shared schema.

Model C 1: Shared database and separate schema

This data model involves housing multiple tenants in the same database, with each tenant having its own set of tables that are grouped into a schema created

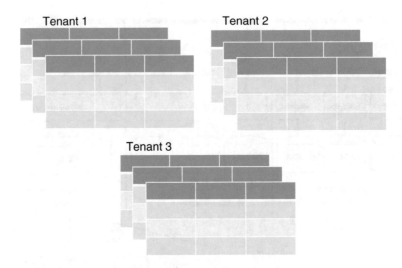

Fig. 1.3 Shared database and separate schema

specifically for the tenant. Figure 1.3 shows this data model. The provisioning database creates a discrete set of tables for the tenant and associates it with the tenant's own schema. Although the tenants' data are in the same database, but with a discrete set of tables, views, stored procedure and triggers. Like the isolated approach, the separate schema approach is relatively easy to implement. This approach offers a moderate degree of logical data isolation for security-conscious tenants, though not as much as a completely isolated system would. It can support a larger number of tenants per database server. A significant drawback of the separate schema approach is that tenant data is harder to restore in the event of a failure. If each tenant has its own database, restoring a single tenant's data means simply restoring the database from the most recent backup. With a separate schema application, restoring the entire database would mean overwriting the data of every tenant on the same database with backup data, regardless of whether each one has experienced any loss or not. Therefore, to restore a single customer's data, the database administrator may have to restore the database to a temporary server, and then import the customer's tables into the production server.

Model C 2: Shared database and shared schema
A second approach involves using the same database and the same schema that is composed of a set of tables to host multiple tenants' data. Figure 1.4 shows this data model. A given table can include records from multiple tenants stored in any order. Therefore, a tenant ID column is added to associates every record with the appropriate tenant.

Of the two approaches explained here, the shared schema approach has the lowest hardware and backup costs, because it allows you to serve the largest number of tenants per database server. However, because multiple tenants share the

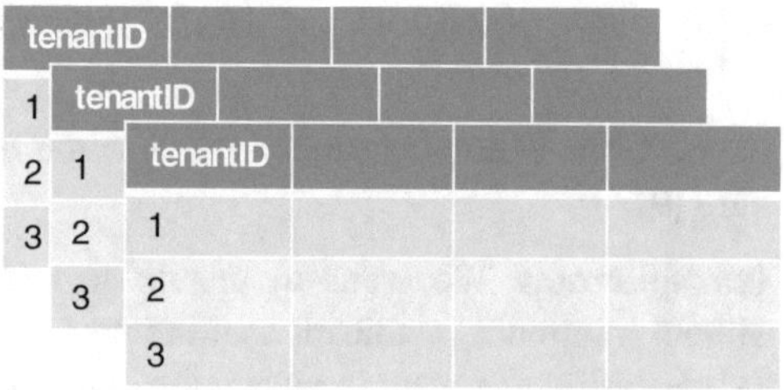

Fig. 1.4 Shared database and shared schema

same database tables, this approach may incur additional development effort in the area of security to ensure that tenants can never access other tenants' data, even in the event of unexpected bugs or attacks. In this context, a multi-tenant data middleware is well designed to optimize and minimize the development work to the utmost. That is the motivation of the proposed multi-tenant data middleware.

1.3   Requirements

1.3.1   Criteria of Multi-tenant Data Middleware

To define what might be called a mature multi-tenant data middleware, we must introduce some additional criteria. From a data architect's point of view, there are three key differentiators that separate a well-designed multi-tenant data middleware from a poorly designed one. A well-designed multi-tenant data middleware is scalable, multi-tenant, and configurable [5].

Scaling the database means maximizing concurrency, and using database resources more efficiently. For example, optimizing data storage, caching reference data, and partitioning large databases are all acceptable ways.

Multi-tenancy may be the most significant paradigm shift that an architect accustomed to designing isolated, single-tenant applications has to make. For example, when a user at one company accesses customer database by using an ERP application service, the application instance that the user connects to may be accommodating users from dozens or even hundreds of other companies. This requires an architecture that maximizes the sharing of data resources across tenants, but that is still able to differentiate data belonging to different customers.

The challenge for the architect is to ensure that the task of configuring databases is simple and easy for the customers, without incurring extra development or operation costs for each configuration. The personalization of tenants' data is permitted using the configuration. For example, the structure of the sharing multi-tenant database might be slightly different.

1.3.2 Requirements of Multi-tenant Data Middleware

We have the following requirements while designing a multi-tenant data middleware [9].

Transparency: We want to enable legacy applications support multi-tenancy without minimum rectification of the source codes.
Extensibility: We want to ensure the extensibility to support the personalization of tenants' data.
Scalability: We want that this middleware has the ability to cache tenants' data to optimize this architecture.
Disaster recovery: We want that the database would be recovered in the case of any unexpected disaster.

1.4 Architecture

In this section, we propose a transparent multi-tenant data middleware, which is shown in Fig. 1.5. This architecture of data middleware is comprised of SQL interceptor, SQL parser, SQL restorer, and SQL router. Some additional components (such as data node and cache) are assisted to optimize this architecture.

Multi-tenant data middleware is a kind of data proxy to support multi-tenancy without the storage of any physical data for the sake of smooth transition. It is transparent to the developers, owning the similar logical set of tables and views as the physical database. Therefore, the legacy application can connect to multi-tenant data middleware transparently. The multi-tenant data middleware provides the logical data isolation for the tenants with higher demand on security.

1.4.1 SQL Interceptor

SQL interceptor is used to intercept the SQLs that are transmitted to SQL parser. An simple implementation of SQL interceptor is using JDBC proxy, which captures all the SQLs in the database driver layer.

1.4.2 SQL Parser

SQL parser is used to parse the fine-grained predicates of SQL statements, such as select predicates, aggregation predicates, where predicates, order predicates, and group predicates.

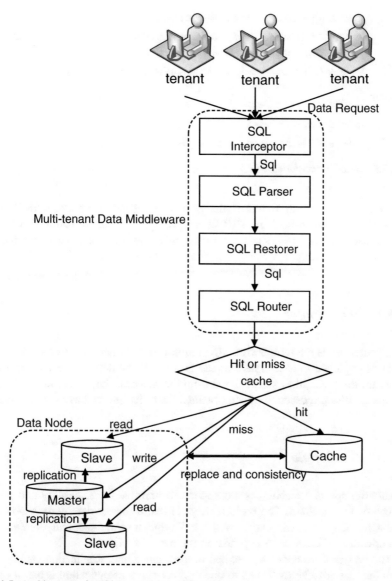

Fig. 1.5 Architecture of multi-tenant data middleware

1.4.3 SQL Restorer

SQL restorer is used to restore the new SQL to the physical sharing data. The new SQL is reorganized from the original SQL predicates and tenantID discriminator. The restoring process is denoted as a mapping: sql(u)->pre(TenantID) $\cup$ T(sql(u), TenantID) $\cup$ post(TenantID), where pre(TenantID) is the pre-personalized

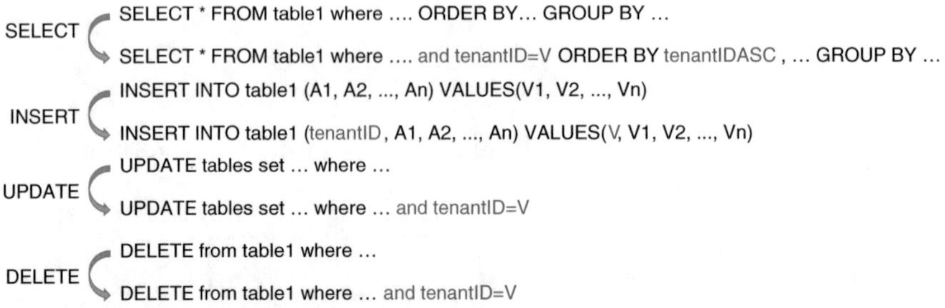

SELECT
⎧ SELECT * FROM table1 where …. ORDER BY… GROUP BY …
⎩ SELECT * FROM table1 where …. and tenantID=V ORDER BY tenantIDASC , … GROUP BY …

INSERT
⎧ INSERT INTO table1 (A1, A2, …, An) VALUES(V1, V2, …, Vn)
⎩ INSERT INTO table1 (tenantID, A1, A2, …, An) VALUES(V, V1, V2, …, Vn)

UPDATE
⎧ UPDATE tables set … where …
⎩ UPDATE tables set … where … and tenantID=V

DELETE
⎧ DELETE from table1 where …
⎩ DELETE from table1 where … and tenantID=V

Fig. 1.6 Restoring transformation function

operation, post(TenantID) means the post-personalized operation, and T is the transformation function. Figure 1.6 is an example of transformation function of select, insert, update, and delete statements, where tenantID indicates the tenant ID column, and V means the value of tenantID in the context.

1.4.4   SQL Router

SQL router sends the reorganized SQL requests to the data node or the cache. The cache is deployed to accelerate the read process. If the data of one column of the query hit the cache, they are obtained from the cache. If the data of one column of the query miss the cache, they are obtained from the master/slave data nodes.

1.4.5   Data Node

Read/write splitting techniques are applied to improve the scalability and performance of the database. The basic concept is that a master data node handles the transactional operations, and slaves handle the non-transactional queries. The identification of transaction depends on the parse of SQLs.

The master/slave nodes are applied in the architecture. Replication enables data from the master to be replicated to one or more slaves. Replication is based on the master server keeping track of all changes to its databases in its binary log. The binary log serves as a written record of all events that modify database structure or data from the moment the server was started. The before and after images are both recorded in the binary log with low impact on the performance of the database. Each slave that connects to the master requests a copy of the binary log. That is, it pulls the data from the master, rather than the master pushing the data to the slave. The slave also executes the events from the binary log that it receives. This has the effect of repeating the original changes just as they were made on the master. Tables

Fig. 1.7 Architecture of the cache

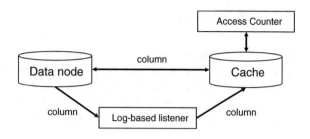

are created or their structure modified, and data is inserted, deleted, and updated according to the changes that were originally made on the master.

1.4.6   Cache

The architecture of the cache is shown in Fig. 1.7, which is a part of the data middleware. In our solution, we adopt the cache to optimize the architecture of the data middleware.

Step 1: Log-based replication from the data node to the cache
The changes of the data node are parsed from the binary log store. This process is called log-based replication in the presence of updates. For a new type of the data node, the only component that needs to change in this architecture is the concrete parser of the binary log. In the case of the insert transactional operation of the data node, it will insert the data of this column into the cache when this column exists in the cache. In the case of the delete transactional operation of the data node, it will delete the corresponding cache when this column exists in the cache. In the case of the update transactional operation of the data node, it will rectify the data in this column in the existence of the cache.

Step 2: Cache listener
One purpose of the access counter is designed to observe the usage of the column in the cache. If the current column access frequency rate is smaller than the average column access frequency rate, the column access frequency count need to be abated. If the subtraction from the current column access frequency rate to the average one descends a negative threshold, we should remove the data in this column in the cache.

Step 3: Cache replacement strategies
If the data in this column of the query hit the cache, it returns the results from the column oriented NoSQL cache. On the contrary, if the data in this column of the query miss the cache, it returns the results from the original data node. If the subtraction from the current column access frequency rate to the average one exceeds a threshold, the data of this column in the data node need to be dynamically translated into the cache. In the case of the hit or miss of the cache, the column access frequency count needs the rectification.

1.4.7   Tenant Context

In order to isolate the tenants' data, tenant ID column is used as a discriminator along with the context session. The legacy application is revised to transfer tenantID discriminator after the success of the users' authentication. For example, if the framework of the application is based on Spring Security, the acquisition of tenantID is configured as a rule of the authentication without any modification of the source codes.

1.5   Evaluation

1.5.1   Cost Analysis

We evaluate the proposed multi-tenant data middleware using cost analysis [10]. The goal is to find the cost-minimal solution for the considered multi-tenant application. Different reengineering measures of varying complexity are necessary for fulfilling this requirement.

This cost of different multi-tenant data models is mainly composed of two major aspects: initial reengineering cost and monthly ongoing cost. The breakeven point of the data model is calculated as:

$$TimetoBreakEven = InitRECosts/\sum MonthlyOngoingCosts$$

where InitRECosts is initial reengineering costs, and MonthlyOngoingCosts means monthly ongoing costs. Therefore, every reengineering activity reducing the monthly ongoing costs will sooner or later be amortized.

The calculation of initial reengineering cost of different data models is as follows.

$$InitCost_{shared} = Cost_{App} + Cost_{Database} + n * (CostApp_{Tenant} + CostDatabase_{Tenant})$$

$$InitCost_{separate} = n * (Cost_{App} + Cost_{Database})$$

$$InitCost_{middleware} = Cost_{App} + Cost_{DataMiddleware} + n * CostApp_{Tenant}$$

The calculation of monthly ongoing cost of different data models is as follows.

$$MonthlyCost_{shared} = MonthlyCost_{Database} + n * (MonthlyCostApp_{Tenant} + MonthlyCostDatabase_{Tenant})$$

$$MonthlyCost_{separate} = n * (MonthlyCost_{Database} + MonthlyCost_{App})$$

$$MonthlyCost_{middleware} = MonthlyCost_{Datamiddleware} + n * MonthlyCostApp_{Tenant}$$

If the service is alive forever, the data model with the lowest incremental monthly ongoing cost is always the best. However, we rather assume that a service

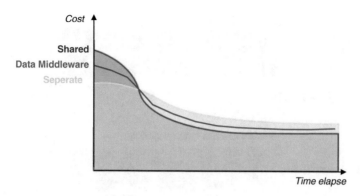

Fig. 1.8 Cost for different multi-tenant data models

gets replaced sooner or later. Thus, the question is whether the time period is sufficiently long to justify huge investments (i.e. service usage time > time to break even).

Figure 1.8 shows the empirical cost of different multi-tenant models. It is indicated that the higher initial reconstruction cost is reasonable for the sake of the low monthly ongoing costs. It is noted that these cost functions in reality might not be linear. Due to the relative complexity of developing a shared architecture, applications that are optimized for a shared approach tend to require a larger development effort than applications that are designed using an isolated approach. The monthly ongoing costs of shared approaches tend to be lower, since they can support more tenants per server. It is shown that our approach using multi-tenant data middleware is a transparent and loose coupled solution for SaaS applications. The cost of our data middleware is between isolated and shared data model at the beginning, but it falls back close to the shared model in the long term.

1.6 Discussion

1.6.1 Extensibility

The requirements of tenant vary from person to person, although they share the similar structure of database. The data middleware is designed to support the personalization of the tenants. There are three approaches to support the personalization [9].

An intuitive approach is using single wide table, which is shown in Fig. 1.9. Single wide table, as its name suggests, stores all the tenant data in the same table with the maximum number of fields. The data model is simply extended to create a preset number of custom fields in every table you wish. This approach often causes the waste of column if the tenant dos not customize this column. This issue is

Single Wide Table

Fig. 1.9 Single wide table

Single wide table with vertical scalability

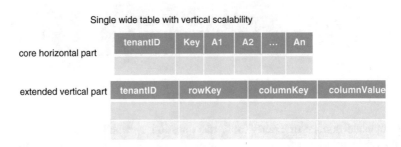

Fig. 1.10 Single wide table with vertical scalability

generally called scheme null issue. In this solution, each tenant that uses at least one or more custom fields gets a row in the combined table, with null fields representing available custom fields that the tenant has not used.

Another improved version of single wide table is single wide table with vertical scalability. This model extracts the personalized data from wide table, and then describes it using extended vertical metadata. Each row in the extended vertical metadata is a key/value pair, which is used to store the personalization of tenants to fulfill the requirements of different tenants. The single wide table with vertical scalability is shown in Fig. 1.10. In the case that the personalization of tenants is identical, the extended vertical metadata can be omitted. The advantage of this approach is that it can reduce the waste of data resources efficiently.

The last approach is multiple wide tables with vertical scalability, which is shown in Fig. 1.11. In the context of multiple wide tables, tenants' data are spread over different single wide tables. That is to say that multiple wide tables with

Multiple wide tables with vertical scalability

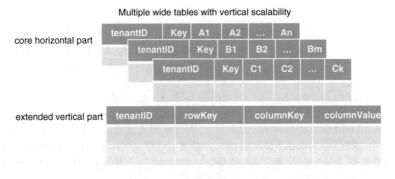

Fig. 1.11 Multiple wide tables with vertical scalability

gradient distribution replace several single wide tables, meeting the demand on dynamic storage requirements of tenants. The tenants' data are determined in either the core horizontal part or the extended vertical part by the requirements on the tenants' customization.

1.6.2   Scalability

The database in the architecture of multi-tenant data middleware is scaled out by partitioning into one or more data nodes according to the key of data. The traditional partitioning algorithm is simple hashing, denoted as PartitioningNumber = Key mod NodeNumber. The shortcoming of this algorithm is that the hash of key would change if the total number of data changed. Consistent Hashing is another revised algorithm, which is based on mapping items to a real angle. Each of the available storage buckets is also pseudo-randomly mapped on to a series of angles around the circle. The bucket where each item should be stored is then chosen by selecting the next highest angle to which an available bucket maps to. The result is that each bucket contains the resources mapping to an angle between it and the next smallest angle.

Some other efforts are also considered to optimize the architecture. When the amount of multi-tenant data grows larger, the reading performance will become a bottleneck. In order to realize the storage of big data, NoSQL techniques are used as the auxiliary mirror storage. Three simple architectures of reading optimization are illustrated as follows [11].

Solution 1: Application-driven dual writes
In this solution, the legacy application writes to the data node and NoSQL mirror in parallel. The advantage of this approach is that it looks simple to implement with the extra development. However, the disadvantage is that both the data node and NoSQL mirror in complete lock step with each other in the face of failures. The data node and NoSQL mirror need to process exactly the similar write operations. Things get even more complicated if the write operations are conditional or have partial update semantics. Figure 1.12 shows this solution.

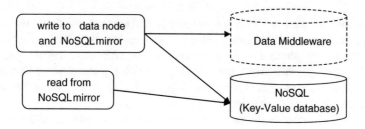

Fig. 1.12 Application-driven dual writes

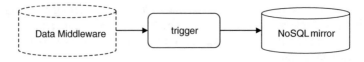

Fig. 1.13 Trigger-based method

Fig. 1.14 Log-based replication method

Solution 2: Trigger-based method

Another method is trigger-based method. Generally, triggers solution log events in another queue collection, and later the data can be extracted to move to the NoSQL. The queue collection might have schema with the following fields: id, table name, row key, timestamp and operation. As an improved form of this approach, some user-defined functions (UDF) of the data node are developed to update the change data without the intermediate queue collection synchronously. Updates to the data node invoke the triggers to either invalidate or refresh the impacted key-value pairs. However, a large amount of triggers or functions will impact the database performance. Figure 1.13 depicts this solution.

Solution 3: Log-based replication method

The last one is log-based replication approach. In this model, the log listener is used to extract changes from its operation logs to move to the NoSQL. This solves the consistency issue, but is practically difficult to implement because the transaction log formats of databases are heterogeneous. This approach is feasible unless the source database naturally supports log events and provides an API for the developers. Figure 1.14 illustrates this solution.

1.6.3 Disaster Recovery

Data backup is the basis of disaster recovery [10]. There are effectively two types of backup, the first is a full backup, either by hot-copying the database files or using dump process to produce SQL to recreate the database at a certain point. The second is the log-based incremental backups, which log every query that changes the data sent to the database server, including DDL statements such as create, alter, and drop tables, and permission statements.

The ideal situation of disaster recovery is to have to restore a minimal amount from the logs. Unfortunately, that is very slow. Therefore, we need frequent full backups and copies of the binary logs since that point. When the disaster happened, restore the database that needs restoring using the full backup and the statements from the logs that pertain directly to the database in question.

In the worst case, when the backup is corrupted, all are not lost even in this case it may be possible to restore just as much data, but it could take a little longer. Because each of our full backups has a full backup SQL, plus extra files with more SQL to bring it to the current state, any given backup is equivalent to the immediately previous backup and the extra files from the backup. That is denoted as: Backup{N} = Backup{N − 1} + LOGS_FROM(Backup{N}). Therefore, if you find that your latest full backup SQL file is corrupt, and then you can substitute the previous full backup SQL file and the previous set of logs to get the same data.

1.7 Conclusions

This chapter aims at proposing a data middleware to support multi-tenancy. The architecture is presented to enable multi-tenancy at the database driver level, which is composed of SQL interceptor, SQL parser, SQL restorer, and SQL router. Furthermore, this data middleware is evaluated to be feasible by the cost. Finally, some features, such as extensibility, scalability and disaster recovery, are discussed.

References

1. Armbrust, M., Fox, A., Griffith, R., Joseph, A. D., Katz, R., Konwinski, A., & Zaharia, M. (2010). A view of cloud computing. *Communications of the ACM, 53*(4), 50–58.
2. Turner, M. J. Worldwide cloud systems management software 2014–2018 forecast. International Data Corporation, research report 247607, 2014.
3. Guo, C. J., Sun, W., Huang, Y., Wang, Z. H., & Gao, B. (2007, July). A framework for native multi-tenancy application development and management. In *The 9th IEEE International Conference on E-Commerce Technology and the 4th IEEE International Conference on Enterprise Computing, E-Commerce, and E-Services, 2007. CEC/EEE 2007* (pp. 551–558). IEEE.
4. Hugos, M. H., & Hulitz, D. (2010). *Business in the cloud: What Every Business Needs to Know About cloud Computing*. NJ: Wiley Publisher.
5. Chong, F., & Carraro, G. (2006). *Architecture strategies for catching the long tail*. MSDN Library, Microsoft Corporation, 9-10.
6. Jacobs, D., & Aulbach, S. (2007). Ruminations on Multi-Tenant Databases. In *BTW* (Vol. 103, pp. 514–521).
7. Ma, K., Yang, B., & Abraham, A. (2012). A template-based model transformation approach for deriving multi-tenant saas applications. *Acta Polytechnica Hungarica, 9*(2), 25–41.
8. Chong, F., Carraro, G., & Wolter, R. (2006). *Multi-tenant data architecture*. MSDN Library, Microsoft Corporation.
9. Ma, K., & Yang, B. (2014). Multiple wide tables with vertical scalability in multitenant sensor cloud systems. *International Journal of Distributed Sensor Networks, 2014*.

10. Ma, K., Chen, Z., Abraham, A., Yang, B., & Sun, R. (2011, October). A transparent data middleware in support of multi-tenancy. In *Next Generation Web Services Practices (NWeSP), 2011 7th International Conference on* (pp. 1–5). IEEE.
11. Hu, Y., & Qu, W. (2013). Efficiently Extracting Change Data from Column Oriented NoSQL Databases. In *Advances in Intelligent Systems and Applications-Volume 2* (pp. 587–598). Springer Berlin Heidelberg.

Chapter 2
Intelligent Web Data Management of NoSQL Data Warehouse

2.1 Introduction

2.1.1 Background

Relational database management systems (RDBMSs) have been widely used for decades to store two dimensional data. However, it will have the I/O bottleneck issues in a world of big data, especially for the read (e.g. query and search) operation. To address this limitation, several systems have already emerged to propose an alternative schema-free database, known as NoSQL [1]. A NoSQL database provides a mechanism for storage and retrieval of data that is modeled in means other than the tabular relations used in relational databases. Motivations for this approach include simplicity of design, horizontal scaling and finer control over availability, especially in using in big data and real-time web applications. The data structure (e.g., tree, graph, and key/value) differs from the RDBMS, and therefore some operations are faster in NoSQL and some in RDBMS. Despite the consistency problem, the benefits of using NoSQL outweigh its disadvantages.

A document-oriented database, or simply called document store for short, is one oof NoSQL databases. It is commonly used to store, retrieve, and manage document-oriented information, which encapsulates and encodes data in some standard formats. Encodings in use include XML, JSON, as well as BSON. MongoDB [2] is an example of the leading document-based NoSQL database, which enables new types of applications, better customer experience, and faster time to market and lower costs for organizations of all sizes. Currently, there are some classical solutions and best practices of the MongoDB.

© Springer International Publishing Switzerland 2016
K. Ma et al., *Intelligent Web Data Management: Software Architectures and Emerging Technologies*, Studies in Computational Intelligence 643, DOI 10.1007/978-3-319-30192-1_2

2.1.2 Challenges and Contributions

Nowadays, data warehouse plays an important part in most of areas of data management including data analysis, data mining and decision support activities in academic and industrial community. Data administrators create the habit of using high-specialized data structures that data warehouses usually maintain materializing their most essential multidimensional perspectives for business analysis and decision. The data in the data warehouse today is enormous, and besides supporting their information needs that they also enrich their common knowledge by bringing new data from very specialized outsourced repositories. The challenge in the data warehouse is that it would have very little effect on the track of the historical data after reducing the data redundancy and compressing the data.

As we know, dimension tables are used to provide subject-oriented information by providing data elements to filter, aggregate or describe facts in a relational data warehouse. Thus, it is quite important that dimensions remain consistent accordingly in some data warehouse's time frame even when some values of their attributes change over time. When this occurs, dimensions that change slowly over time rather than changing on regular schedule. It is often called as slowly changing dimensions (SCD) [3, 4].

However, most of the publications just discuss the SCD approaches of RDBMS. In the field of the emerging NoSQL databases, few publications have mentioned the technologies about how to formulate the data warehouse of NoSQL. Not only the structure of NoSQL is different from RDBMS, but also NoSQL instance has bigger data than RDBMS in practice. It is just natural that whether distributed computing can integrate with the SCD approach or not is wondered. Currently, MapReduce is a functional programming model for processing large data sets with a parallel, distributed algorithm on a cluster. Therefore, MapReduce framework is used to accelerate the formulation of the data warehouse.

In the rest of this chapter, a SCD approach of document-based NoSQL data warehouse is introduced. MapReduce is taken advantage of to benefit SCD in the scenario of NoSQL databases effectively. The main difficulty to construct document-based NoSQL data warehouses is to achieve the balance among the complexity, efficiency and flexibility. The contributions of this chapter are into several folds. First, this SCD approach can highly compress the historical data in the data warehouse of schema-free document stores without data redundancy. Second, this SCD approach can keep track of the history, providing quicker access to the snapshot at every point of a day or over a period of time. Finally, it is managed to assure that the formulation of the daily cell with an effective lifecycle tag is efficient and transparent to the business of applications.

2.2   Related Works and Emerging Techniques

Although the structure of the NoSQL is different from RDBMS, a relational database is a special case of NoSQL in a broad way. The SCD principles to RDBMS adapt to NoSQL through continuous innovative improvement. Currently, there are at least three categories of strategies to the SCD. The first approach is daily full backup. In this solution, all the daily backups that stand alone formulate the whole data warehouse, which are simple but need very large storage spaces. Each backup is a snapshot of the database at a particular point in time. The second approach is daily incremental backup. An incremental backup includes only those things that have changed since the previous backup and saves those things into a separate and additional location. By definition, the first incremental backup is a full backup since it backs up everything since there is no previous backup to compare to. The next incremental backup backs up only those files that have changed since the previous backup was taken. This incremental backup can result in a much smaller backup. The cost of using incremental backups is one of the management. Since each incremental backup relies on the backup that preceded it, in order to restore the database to an arbitrary point in time all the incremental backups must be available to perform the restore. The pros of this SCD approach cost significantly less disk space used compared to an equivalent set of full backups. Moreover, The cons of this SCD approach are that the baseline full backup and all the incremental backups must be preserved and available in order to restore. Besides, the deleted historical data are not easy to be tracked. Thus, another improvement of this approach is daily incremental backup with the invalid partition. The content of the data warehouse is divided into the last full backup, this daily latest increment and this daily invalid partition. For the track of the deleted data, this is got from the daily invalid partition. The last approach is daily differential backup. Differential backup is a kind of hybrid approach. In reality, it is just incremental backup, but with a fixed starting point. Rather than backing up only changes from the previous day, each differential backup includes all the changes from the baseline of full backup. Compared with the second incremental approach, each day can be restored from only two backups, the initial baseline full plus that day's differential.

Although SCD is not a new problem during the dimensional modeling phase of a data warehouse, the state of art of SCD in the context of document-oriented NoSQL databases is ambiguous. There are many mature SCD approaches to make data warehouse of specific RDBMS. As for the emerging schema-free NoSQL databases, few publications have mentioned the technologies about how to formulate the data warehouse.

2.2.1   Slowly Changing Dimensions of RDBMS

While dimension table attributes are relatively static, they are not fixed forever. Dimension attributes change, albeit rather slowly, over time. Dimensional designers must engage business representatives proactively to help determine the appropriate change-handling strategy. While it is assumed that accurate change tracking is unnecessary, business users may be assumed that the data warehouse will allow them to see the impact of each and every dimension change. Special oriented Extract-Transform-Load (ETL) processes are responsible to maintain SCD, acting accordingly to the updating strategy previously defined. Usually, a Change Data Capture (CDC) process detects the update event and records it accordingly to some predefined SCD requisites. Although it is always spoken in terms of process, SCD strategies are applied at the attribute level, for a particular dimension table. For each dimension table, designers define what kind of update attributes will be modified over time. They must prevent all the possible cases of attribute changing, once it is not very recommendable to change dimension table structures when the DWS is already in production. Over a same SCD, distinct updating strategies for a single record is to be applied. Frequently, this involves different processes for different SCD strategies.

Dealing with the issues involves SCD management methodologies referred to as Type 0 through 6 [5]. Type 6 SCDs are also sometimes called Hybrid SCDs.

Type 0: The passive method
The Type 0 method is passive. It manages dimensional changes and no action is performed. Some dimension data can remain the same as it was first time inserted, others may be overwritten. In certain circumstances history is preserved with a Type 0. High order types are employed to guarantee the preservation of history whereas Type 0 provides the least or no control.

Type 1: Overwriting the old value
In this method no history of dimension changes is kept in the database. The old dimension value is simply overwritten be the new one. Thus, it does not track historical data. This type is easy to maintain and is often use for data which changes are caused by processing corrections (e.g. removal special characters, correcting spelling errors).

An example of Type 1 is shown in Fig. 2.1. This example is a teacher table with attributes ID, number, name, and title. ID is the natural key, and number is a surrogate key. If the teacher is promote to professor, the record would be over-written. The disadvantage of the Type 1 method is that there is no history in the data warehouse. It has the advantage however that it's easy to maintain.

Type 2: Adding a dimension row
In this methodology, all history of dimension changes is kept by creating multiple records for a given natural key in the dimensional tables. Unlimited history is preserved for each insert. Both the prior and new rows contain as attributes the natural key. The first method to track the data is called 'current indicator'. There

ID	number	name	title
1	0001	Jim Green	lecturer

BEFORE

ID	number	name	title
1	0001	Jim Green	professor

AFTER

Fig. 2.1 An example of type 1

could be only one record with current indicator set to 'Y'. Another method to track the data is called 'effective date'. For 'effective date' columns, i.e. start_date and end_date, the end_date for current record usually is set to value 9999-12-31. The 9999-12-31 end_date in row two indicates the current record version. Two examples of these two methods are shown in Figs. 2.2 and 2.3. Introducing changes to the dimensional model in type 2 could be very expensive database operation so it is not recommended to use it in dimensions where a new attribute could be added in the future.

Transactions that reference a particular surrogate key are then permanently bound to the time slices defined by that row of the slowly changing dimension table. An aggregate table summarizing facts by state continues to reflect the historical state. If there are retrospective changes made to the contents of the dimension, or if new attributes are added to the dimension which have different

BEFORE

ID	number	name	title	current
1	0001	Jim Green	lecturer	Y

AFTER

ID	number	name	title	current
1	0001	Jim Green	lecturer	N
2	0001	Jim Green	professor	Y

Fig. 2.2 Current indicator method of type 2

BEFORE

ID	number	name	title	start_date	end_date
1	0001	Jim Green	lecturer	1999-01-01	9999-12-31

AFTER

ID	number	name	title	start_date	end_date
1	0001	Jim Green	lecturer	1999-01-01	2013-12-31
2	0001	Jim Green	professor	2014-01-01	9999-12-31

Fig. 2.3 Effective date method of type 2

effective dates from those already defined, then this can result in the existing transactions needing to be updated to reflect the new situation. This can be an expensive database operation, so Type 2 SCDs are not a good choice if the dimensional model is subject to change.

Type 3: Adding a dimension column
This method tracks changes using separate columns and preserves limited history. The Type 3 preserves limited history as it is limited to the number of columns designated for storing historical data. The original table structure in Type 1 and Type 2 is the same but Type 3 adds additional columns. In the following example shown in Fig. 2.4, an additional column has been added to the table to record the title of a teacher—only the previous history is stored. The new value is loaded into 'current' column and the old one into 'previous' column. This record contains a column for the previous title and current title track the changes. Figure 2.4 shows an example of Type 3.

Type 4: Using historical table
The Type 4 method is usually referred to as using historical table, where one table keeps the current data, and an additional table is used to keep a record of some or all changes. Both the surrogate keys are referenced in the fact table to enhance query performance. For the above example the original table name is teacher and the history table is teacher_history. Figure 2.5 shows an example of Type 4. This method resembles how database audit tables and change data capture techniques function.

Fig. 2.4 An example of type 3

BEFORE

ID	number	name	previous_title	current_title
1	0001	Jim Green	lecturer	lecturer

AFTER

ID	number	name	previous_title	current_title
1	0001	Jim Green	lecturer	lecturer
2	0001	Jim Green	lecturer	professor

Fig. 2.5 An example of type 4

teacher

ID	number	name	title
1	0001	Jim Green	professor

teacher_history

ID	number	name	title	create_date
1	0001	Jim Green	lecturer	2013-12-31

ID	number	name	previous_title	current_title	start_date	end_date	current

BEFORE

ID	number	name	previous_title	current_title	start_date	end_date	current
1	0001	Jim Green	lecturer	lecturer	1999-01-01	9999-12-31	Y

AFTER

ID	number	name	previous_title	current_title	start_date	end_date	current
1	0001	Jim Green	lecturer	lecturer	1999-01-01	2013-12-31	N
2	0001	Jim Green	lecturer	professor	2014-01-01	9999-12-31	Y

Fig. 2.6 An example of type 6

Type 6: Hybrid methods of types 1, 2, 3 (1 + 2 + 3 = 6)
The Type 6 method combines the approaches of types 1, 2 and 3 (1 + 2 + 3 = 6). This hybrid method is also called unpredictable changes with single-version overlay. Figure 2.6 shows an example of Type 6.

However, the above classical SCD approaches have some limitations. Type 1 cannot keep track of historical data. Type 2, 3 and 4 have too much redundant data. In our opinion, keeping the track of historical data is the unique SCD question, especially for the NoSQL databases. In our point of view, when keeping history in dimension data is spoken about, all types of SCD can be implemented quite effectively using only SCD type 4. Classifying an SCD as a new Type X is not exactly true, types of SCD are applied to attributes, so in general dimensions with attributes that are type 1 are considered, other attributes that are type 2 or type 3.

2.2.2   Slowly Changing Dimensions of NoSQL

Since the data in the data warehouse are used for data analysis with SCDs, we are only concerned about the daily data rather than the real-time data. That is to say that only the last change of the data each day is valid. Thus, we will merge all the real-time data each day into a set of final daily data. There are two solutions to slowly changing dimensions of NoSQL [6].

Solution 1: document with timestamp
Obviously, we can conclude that the historical collection dimension is a feasible SCD solution to the schema-free document stores. In this solution, the historical data are saved in another separate collection with the timestamp every day. In order to make further comparison, we call this approach "document with timestamp" next. Figure 2.7 shows the structure of document with timestamp. However, this solution has too many disadvantages. The first one is that the data warehouse is

Fig. 2.7 Document with timestamp

ID	A1	...	An	timestamp

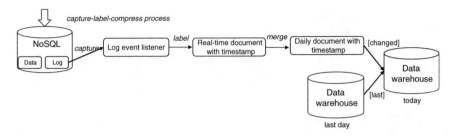

Fig. 2.8 Process of the formulation of the NoSQL data warehouse using document with timestamp

huge enough so that the storage becomes a disaster over a long period. The second one is that it will consume too many excessive storage spaces due to the redundant data. The process of the formulation of the NoSQL data warehouse using this solution is shown in Fig. 2.8. First, we capture the changed document-based data from the NoSQL using the log event listener. Second, we merge the changed data with a timestamp. Third, we compress all the real-time documents with a timestamp on a day with the same surrogate key into one document. At last, all the compressed daily documents with timestamp plus the data warehouse last day formulate the new data warehouse today.

Solution 2: document with a lifecycle tag

Another improved version of the document with timestamp solution is called "document with a lifecycle tag". Figure 2.9 shows the structure of document with a lifecycle tag. In this solution, we use the start and end timestamp instead of the unique timestamp. Although this solution saves the storage space compared with the first solution, it has the challenge on how to generate the collection dimension in the data warehouse. Similar to the timestamp solution, the process of the formulation of the NoSQL data warehouse using this solution is shown in Fig. 2.10. First, we capture the changed document-based data from the NoSQL using the log event listener. Second, we label the changed data with lifecycle tags. Third, we compress all the real-time documents with lifecycle tags on a day with the same surrogate key into one document. At last, all the compressed daily documents with lifecycle tags plus the data warehouse last day formulate the new data warehouse today.

The above two solutions have the same disadvantage that the daily documents in the data warehouse have excessive storage space due to the data redundancy. For this purpose, we design a new SCD approach adapted to the schema-free document stores using fine-grained element. We call this approach "cell with an effective

Fig. 2.9 Document with a lifecycle tag

ID	A1	...	An	start_date	end_date

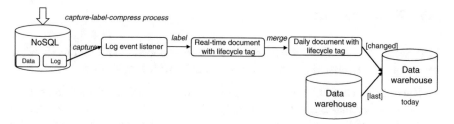

Fig. 2.10 Process of the formulation of the NoSQL data warehouse using document with a lifecycle tag

lifecycle tag using MapReduce". This idea comes on this premise that the changing of the database is slow enough to have redundant data in the data warehouse.

2.2.3   MapReduce Framework

MapReduce is a programming model for parallel processing large data sets, which was proposed by Google in 2004 [7]. A popular free implementation is Apache Hadoop. The model is inspired by the map and reduce functions commonly used in functional programming, although their purpose in the MapReduce framework is not the same as their original forms. The process is divided into two steps: map and reduce. Developers specify Map functions which take an input pair and produces a set of intermediate key/value pairs. The MapReduce library then groups together all intermediate values associated with the same intermediate keys, and passes them to the Reduce functions which are also specified by the programmers; The Reduce function accepts an intermediate key and a set of values for that key. It merges together these values to form a possibly smaller set of values. The intermediate values are supplied to the user's reduce function via an iterator. The programming model of MapReduce framework is shown as follows:

- map: (k1, v1) -> list(k2, v2). Map Function. Map function takes an input key-value pair (k1, v1) and produces a set of intermediate key-value pairs (k2, v2).
- reduce: (k2, list(v2)) -> list(k3, v3). Reduce Function. Reduce function merges the intermediate key-value pairs (k2, list(v2)) together to produce a smaller set of values list(v3).

As for the application scenarios of SCD, MapReduce can improve the computational capacity to the creation of the NoSQL data warehouse. Map function in MapReduce corresponds with mapping the document-based data into real-time cells with lifecycle tags, and reduce function in MapReduce works in concert with merge all the real-time cells with lifecycle tags into daily cells with lifecycle tags.

2.3   Requirements

First, we have the following requirements and their corresponding solutions while designing slowly changing dimensions of NoSQL with MapReduce.

- **Keeping track of history**: Since the historical data are used for data analysis and decision making, we want to keeping track of history. In our solution, we use cells with an effective lifecycle tag using MapReduce to track all the changes to the source NoSQL database.
- **Quick access to the historical snapshot at every point of a day or over a period of time**: We want to further analyze the data in the NoSQL data warehouse. This implies needing to support the quick access to the historical data on a specific day or over a period of time.
- **Reduce the storage space and remove the redundant data from the data warehouse**: Since the data in the data warehouse are compressed as the final daily cells with effective lifecycle tags, we expect to reduce the storage space and remove the redundant data of the data warehouse. That is to say that the current cell with timestamp and lifecycle tag solutions need the high compression ratio of the NoSQL data warehouse. In our solution, we split the changed document stores into sets of cells with effective lifecycle tag divided by column.
- **The high efficiency of the creation**: Due to the large amount of data in the source NoSQL database, the creation of the NoSQL data warehouse can benefit from the distributed computing. Therefore, we use MapReduce framework to generate the schema-free document stores in the data warehouse.
- **Data consistency preservation**: We want to preserve the consistency semantics that the data warehouse provides. Since the process of data analysis and decision making need not be done in real time, we can tolerate the delay of the data consistency. We can miss the latest but cannot provide the incorrect historical data.
- **Transparency to the business**: We expect the creation of the data warehouse is independent of the application using NoSQL databases. This will be implemented with the change data capture engine independent of the source database. Thus, we design a log-based change data capture that will impact the performance of the source database at a minimum.

2.4   Architecture

In this section, we discuss the proposed architecture for changing the dimensions of NoSQL with MapReduce.

2.4.1 Deployment Architecture

The typical architecture of slowly changing dimensions of NoSQL with MapReduce is shown in Fig. 2.11. The architecture is composed of four layers in order to reduce the complexity. From the top to the bottom, they are data layer, computing layer, historical storage layer and application layer respectively.

On top is the data layer. Since EDSM of schema-free document stores is designed, MongoDB is adopted as the NoSQL in this layer. All the historical data in the data warehouse come from this source NoSQL. The second layer from the top is the computing layer, which is the core of our proposed middleware. The Capture-Map-Reduce procedure is used to generate the data in the data warehouse. The details are discussed in the next section. The second layer from the bottom is the historical storage layer. The historical data in this layer are stored. At the bottom is the application layer. This layer consists of all kinds of applications using the data warehouse, such as decision making, data analysis and data mining.

2.4.2 Capture-Map-Reduce Procedure

As the name suggests, the tentative plan for the capture-map-reduce procedure consists of the following steps: capture, map and reduce procedure. It is shown in Fig. 2.12. Strictly speaking, the capture-map-reduce procedure refers to a kind of ETL.

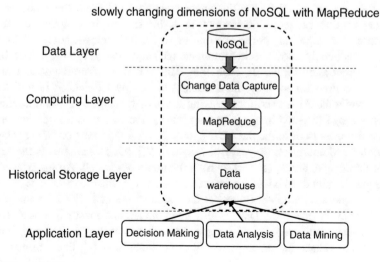

Fig. 2.11 Architecture

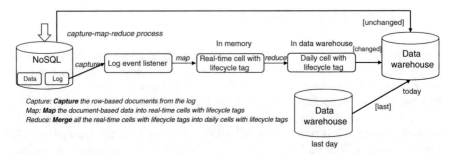

Fig. 2.12 Capture-map-reduce process

2.4.3   Log-Based Capture

Our log-based engine has been designed to support the extraction from different source NoSQL databases. This flexibility is achieved by allowing different captures to be developed and plugged into the capture engine. In this Chapter, schema-free document stores are mainly discussed. This capturer is typically an embedded thread and must follow a few semantic constraints. For example, oplog (operations log) is used to implement MongoDB capture.

In the scenario of change data capture (CDC), only the changed data reading from the operational log are concerned. The document-oriented NoSQL products often provide document-based replication (DBR) events to capture changed data from the operational log. This API is used to connect the listener to get the incremental data. By interpreting the contents of the database operational log one can capture the changes made to the database in a non-intrusive manner.

In the process of DBR, the document events parsed from the oplog contain changes to one or more documents in a transaction in a given collection. A document change may consist of one or two full document copies in its turn. These are generally known as before image (BI) and after image (AI). Each image has a different purpose: the BI, containing data as it was before the document was modified, is used for locating the document to be updated/deleted in the storage engine, while the AI is used for replaying the actual changes. In addition, the one and only usage for such image is to help finding the correct cell to be updated or deleted. In order to describe the CDC approach, the representation of the image is given, denoted as image: (keyname, k_1, k_2, ..., k_n), where keyname is the unique key of this image, and k_t ($1 \Leftarrow t \Leftarrow n$) is the name of the cell. An instance of this image is generally denoted as (keyvalue, v_1, v_2, ..., v_n), where keyvalue is the value of this unique key, v_t ($1 \Leftarrow t \Leftarrow n$) is the value of the cell. Not both images are needed for every operation. Deletes only need the BI, and inserts just need the AI, while updates generate pairs of images for each row changed. Summing up, BI must contain values that uniquely identifies documents, acting like a primary key, while AI must contain values that make possible changing the document according to the original execution.

Most databases have binary logs that contain the log of changes as they are applied to the database. However, it is fragile to mine these logs and reverse-engineer the structure, because there is no guarantee that the format will be stable across multiple subsequent versions. In the case of MongoDB though, it is possible to tap into the storage engine API. MongoDB product itself provides a stable that has been used to build many commercial and open-source storage engines. The capture interface is implemented in the form of adapters so that it can be reused to support more NoSQL databases.

2.4.4 MapReduce

In the MapReduce procedure, the changed data (in the form of document-based data) from the log event listener are transformed into the final daily cells with lifecycle tags in the data warehouse. The innovation of our approach is utilizing MapReduce framework. In the first Map procedure, the document-based data are mapped into real-time cells with effective lifecycle tags. For the insert changes, they are mapped into the newborn cells with effective lifecycle tag. For the delete changes, thay are mapped into the dead cells with effective lifecycle tag. For the update changes, they are mapped into the dead and newborn cells with effective lifecycle tags in order. The Map procedure is shown in Fig. 2.13, where UUID and

---

Algorithm 1 Map algorithm *map*

Input:
    Collection name *source*;
Output:
    cells with effective lifecycle tag *cells*;
1: **procedure** map()
2: **switch** (*operation*) //log event
3: **case insert:**
4:    // $afterImage \leftarrow (keyname, k_1, k_2, ..., k_n) : (keyvalue, v_1, v_2, ..., v_n)$;
5:    Cell.add(($naturalKey, keyname, keyname, tag) : (Cell.UUID(), keyvalue, keyvalue, (currentstamp, null))$));
6:    **for** $i = 1$ *to* n **do**
7:        Cell.add(($naturalKey, keyname, k_i, tag) : (Cell.UUID(), keyvalue, v_i, (currentstamp, null))$));
8:    **end for**
9: **case delete:**
10: $beforeImage \leftarrow (keyname, k_1, k_2, ..., k_n) : (keyvalue, v_1, v_2, ..., v_n)$;
11: **for** $i = 1$ *to* n **do**
12:    Cell.edit(($naturalKey, keyname, k_i, tag) : (Cell.lastUUID(), keyvalue, v_i, (lastStart(), currentstamp))$));
13: **end for**
14: **case update:**
15: $beforeImage \leftarrow (keyname, k_1, k_2, ..., k_n) : (keyvalue, b_1, b_2, ..., b_n)$ from *pair*;
16: $afterImage \leftarrow (keyname, k_1, k_2, ..., k_n) : (keyvalue, a_1, a_2, ..., a_n)$ from *pair*;
17: **for** $i = 1$ *to* n **do**
18:    Cell.edit(($naturalKey, keyname, k_i, tag) : (Cell.lastUUID(), keyvalue, b_i, (Cell.lastStart(), currentstamp))$));
19:    Cell.add(($naturalKey, keyname, k_i, tag) : (Cell.UUID(), keyvalue, a_i, (currentstamp, null))$));
20: **end for**
21: **end switch**
22: **endprocedure**

Fig. 2.13 Map algorithm

Algorithm 2 Reduce algorithm *reduce*

Input:
    real-time cells with effective lifecycle tags *cells*;
Output:
    daily documents *documents*;
1: **procedure** reduce()
2: classify *cells* by surrogate key;
3: **for each** *surrogateKey* $\in$ *cells* **do**
4:    list=the cells including this *surrogateKey*;
5:    merge the cells *list* into a cell *newcell* with the last value today;
6: **end for**
7: **endprocedure**

Fig. 2.14 Reduce algorithm

lastUUID are the functions to obtain an unique key and the last unique key of a cell respectively, and lastStart is the function to get the last start timestamp of the cell. Besides, the add and edit is the function to insert and update the cell.

In the second Reduce procedure, all the real-time cells with effective lifecycle tags are merged into daily cells with effective lifecycle tags. The Reduce procedure is shown in Fig. 2.14. First, all the cells with effective lifecycle tags within a day are collected. Next, they are classified by the surrogate key. Since the daily data rather than the real-time data for further analysis are concerned, the set of the cells with the same surrogate key is merged. The value of this cell depends on the last value of this day.

2.5   Evaluation

Slowly changing dimension (SCD) problem applies to cases where the attribute of a record varies over time. In fact, most of the business data in the NoSQL applications are of this kind. Since only a small part of the cell and the document varies over time, this generates a lot of redundant data. In the SCD process, the probability of the changing of a document and cell is very small (generally less than 10 %). To test the performance of different NoSQL SCD solutions, source schema-free document stores are generated using the script we design. This generator was configured to create results for 180 days each having average 5000 documents. Thus, data for one month gave 900,000 fact documents. Our cell with an effective lifecycle tag solution is compared with the current document with timestamp and lifecycle tag solutions. Our experiments were performed on a six-core server (Xeon(R) CPU

E7-4807 @1.87 GHz, 130 G RAM, 500TB SATA-III disk, Gigabit Ethernet) running MongoDB 2.2.7 and Hadoop MapReduce 2.2. The system was configured with a Windows Server 2012 x64.

The source NoSQL and its data warehouse are initialized as empty. After that, the script is executed to generate the data in the source document stores. At the same time. Our EDSM begin to formulate the data in the data warehouse. The experimental data of document with timestamp and lifecycle tag solutions are observed, and our cell with an effective lifecycle tag solution. In the scenario of SCD, it is assumed that a document remains unchanged with probability 90 %, and a key in a document remains unchanged with probability 80 %. While in the scenario of non-SCD, it is assumed that a document remains unchanged with probability 40 %, and a key in a document remains unchanged with probability 50 %.

2.5.1   Redundancy Rate

First, the redundancy experiment of SCD and non-SCD solution to illustrate the advantage of our solution have been made. Figures 2.15 and 2.16 show the

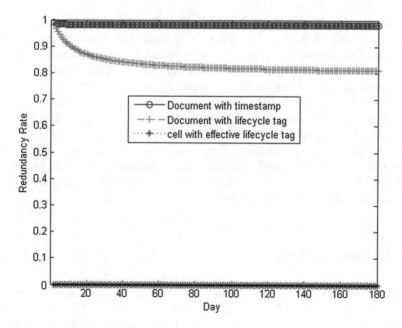

Fig. 2.15 Redundancy rate of SCD

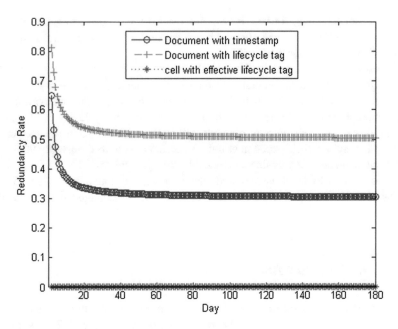

Fig. 2.16 Redundancy rate of non-SCD

redundancy rate every day in the scenario of SCD and non-SCD respectively. As depicted in SCD Fig. 2.15, the first two solutions have the serious bottleneck problem of the redundancy. The redundancy rate of the document with timestamp solution heads toward 100 % in the SCD scenario, since all the documents need to be stored regardless of whether the documents are changed or not. The redundancy rate of the document with lifecycle tag solution is diminished gradually due to the consolidation of the unchanged data at different days. As depicted in non-SCD Fig. 2.16, the redundancy rate of the document with timestamp solution is down with the unchanged probability of the documents. No matter the SCD and non-SCD, our cell with an effective lifecycle tag solution handles the redundancy problem better. Owing to the breakup of the documents into cells, our solution produced no additional redundant data.

2.5.2 Storage Space

Next, the storage space of different solutions is analyzed. It is assumed that the data size increases by average 5000 documents from the source NoSQL every day. The storage space of the corresponding data warehouse every day is observed. Figures 2.17 and 2.18 shows the variation of different three solutions respectively.

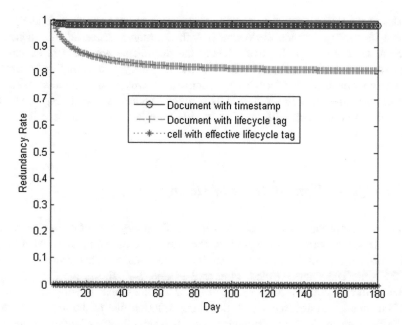

Fig. 2.17 Storage space of SCD

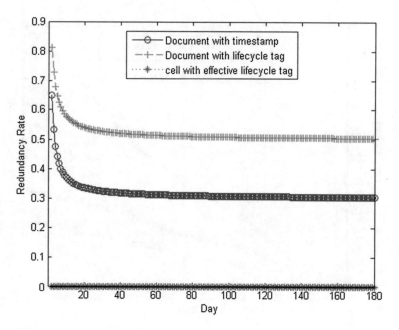

Fig. 2.18 Storage space of non-SCD

Scales of the data size of the corresponding data warehouse increase linearly. The fastest increasing solution is document with timestamp, since all the documents need to be stored regardless of whether the documents are changed or not. As depicted in SCD Fig. 2.17, our cell with effective lifecycle tag solution is superior to the document with lifecycle tag solution. However, our solution takes no remarkable superiority in the non-SCD scenery, which is shown in Fig. 2.18. It is concluded that our cell with an effective lifecycle tag solution is effective in the SCD scenario.

2.5.3   Query Time of Track of History

In the process of the storage space experiment, the query time of the historical data in the data warehouse is measured. In order to make the comparison of different solutions, the historical data on the first day are selected. 8 points to record the time of the same query are selected. Figures 2.19 and 2.20 shows the query time of different three solutions. As depicted in Fig. 2.19, the worst showing on the query performance is the document with timestamp solution due to the large amount of historical data in the data warehouse. The increase of our cell with an effective lifecycle tag solution is lower. Next, the query time over a period of time is

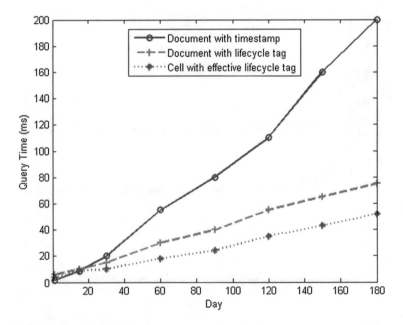

Fig. 2.19 Query time of SCD

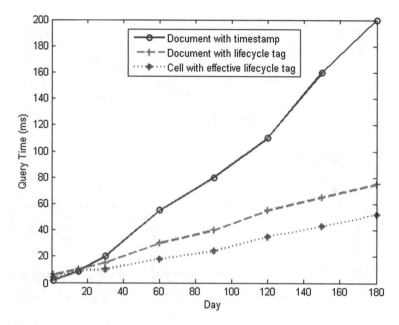

Fig. 2.20   Query time of non-SCD

measured. The historical data from the 30 to 40 days are selected during the 180 days. As depicted in Fig. 2.20, the query time is always the lowest. It is concluded that our solution is feasible in practice.

Generally, data redundancy and data consistency conflict with each other. slowly changing dimensions of NoSQL with MapReduce attempts to assure low data redundancy as well as eventual and weak consistency. The data in the data warehouse might miss the latest value. One day delay is tolerated, since data warehouse is used in the field of data analysis and decision making.

2.5.4   Execution Time of Creation

Finally, the execution time of label-merge procedure (used by the first two solutions) is compared with map-reduce procedure that we use. As depicted in Fig. 2.21, map-reduce solution consumes the shortest time with the deployed Hadoop MapReduce framework.

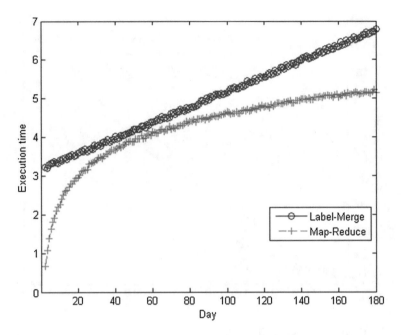

Fig. 2.21 Execution time of label-merge and map-reduce procedures

2.6 Discussion

In this section, we discuss the structure of cell with effective lifecycle tag, and the extreme data storage principles for optimization.

In order to define the slowly changing dimensions of NoSQL with MapReduce, a mathematical representation of the fine-grained model is necessary. In our solution, we map the changes to source NoSQL database into sets of real-time cells with effective lifecycle tags divided by column. Then the real-time cells are merged into daily cells. We assume that the Extract-Transformation-Load (ETL) process extracts all the changed data exists in a source NoSQL database to load into a specific dimension of the data warehouse.

2.6.1 Effective Lifecycle Tag

First, we give the definition of the effective lifecycle tag. The effective lifecycle tag is a 2-tuple of start and end element, which implies the lifecycle of a cell in the NoSQL data warehouse. The effective lifecycle tag is denoted as $(start, end)$. The first element of an effective lifecycle tag is the start timestamp, and the second element of an effective lifecycle tag indicates the end timestamp.

Next, we define some special effective lifecycle tags.

- **Newborn effective lifecycle tag**: For the inserted document of the source NoSQL database, we will split it into several cells in the data warehouse by column. Its value of the effective lifecycle tag is (currentTimestamp, null), where currentTimestamp means the current timestamp.
- **Killed effective lifecycle tag**: For the deleted document of the source NoSQL database, we will split it into several cells in the data warehouse by column. Its value of the effective lifecycle tag is (start, currentTimestamp), where currentTimestamp means the current timestamp.
- **Active effective lifecycle tag**: When the end tag of the effective lifecycle tag is null, it indicates that this cell is alive.
- **Dead effective lifecycle tag**: When the end tag of the effective lifecycle tag is less than the current timestamp, it indicates that this cell becomes the historical data.

An update operation of the source NoSQL database is split into two atomic operations (delete the original data, and insert the new data). Seen from the dimensional point of view, update means the death of the original data and the born of the new data.

2.6.2   Cell with Effective Lifecycle Tag

We define the cell with effective lifecycle tag in the data warehouse as a list of attributes that comprise a natural key, a surrogate key, a regular key and an effective lifecycle tag, denoted as (naturalKey, surrogateKey, regularKey, tag), where naturalKey is the unique identification of the dimension of the data warehouse, surrogateKey is the unique identification of the fact of the source NoSQL database, regularKey is the name of the attribute, and tag is the effective lifecycle tag introduced in the last section. This definition is the metamodel of the cell with an effective lifecycle tag. That is to say that the cell in practice is the instance of this model.

2.6.3   Extreme Data Storage Principles

From the above characteristics of the cell with an effective lifecycle tag, we arrive at the following conclusions.

- Each changed document in the source NoSQL database generates several cells with effective lifecycle tags.
- Each cell has one and only one effective lifecycle tag.
- Each effective lifecycle tag belongs to a set of cells.

In fact, the data administrators only pay close attention to the daily data of the data warehouse rather than the real-time data. Therefore, we design the extreme data storage middleware to further compress the real-time data into the daily data. In other words, the timestamp of the effective lifecycle tag is exact to the day. We will merge all the real-time cells with effective lifecycle tag into daily cells with a lifecycle tag as the extreme compression using MapReduce framework. This will be discussed in the next section.

Next, we analyze the principles of the historical snapshot. Figure 2.22 shows the method to get the historical 1-day snapshot. The cells with effective lifecycle tags

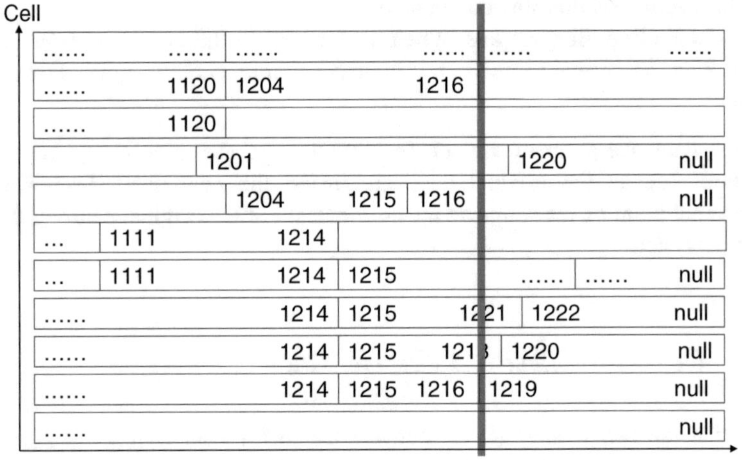

Cell

......	......	......		......	......		......
......	1120	1204		1216			
......	1120						
		1201			1220		null
		1204	1215	1216			null
...	1111	1214					
...	1111	1214	1215		......	......	null
......		1214	1215	1221	1222		null
......		1214	1215	121	1220		null
......		1214	1215	1216	1219		null
......							null

Fig. 2.22 The method to get 1-day historical snapshot

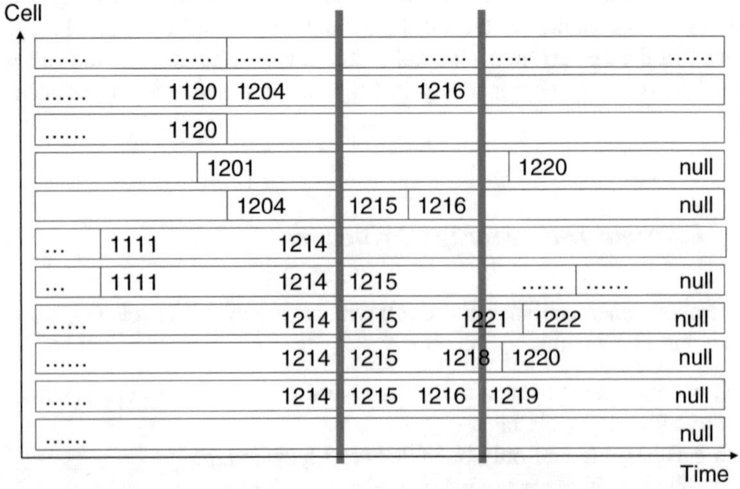

Cell

......	......	......		......	......		......
......	1120	1204		1216			
......	1120						
		1201			1220		null
		1204	1215	1216			null
...	1111	1214					
...	1111	1214	1215		......	......	null
......		1214	1215	1221	1222		null
......		1214	1215	1218	1220		null
......		1214	1215	1216	1219		null
......							null

Time

Fig. 2.23 The method to get the historical snapshot over a period of time

that are penetrated by the black line formulate the full snapshot of the historical data in the date 1217. The snapshot of any day is achieved is this way.

Figure 2.23 shows the method to get the historical snapshot over a period of time. The cells with effective lifecycle tag that are penetrated by the two black lines formulate the full snapshot of the historical data between the date 1215 and 1217. The snapshot for some time is achieved in this way.

2.7   Conclusions

As the emerging NoSQL databases, MongoDB is a document database that provides high performance, high availability, and easy scalability. Currently, there are more and more NoSQL applications, such as the social networking site. However, it lacks the approaches to the NoSQL data warehouse. In this chapter, a methodology for slowly changing dimensions of NoSQL with MapReduce is introduced. The innovation of this method is that it reduces the data redundancy in the data warehouse, and optimize the storage structure. Due to the support of the parallel and distributed computing framework, MapReduce is used to improve the performance of the creation of the NoSQL data warehouse.

References

1. Cattell, R. (2011). Scalable SQL and NoSQL data stores. *ACM SIGMOD Record, 39*(4), 12–27.
2. Chodorow, K. (2013). *MongoDB: The definitive guide*. O'Reilly Media, Inc.
3. Kimball, R., Ross, M., Thornthwaite, W., Mundy, J., & Becker, B. (2008). The data warehouse lifecycle toolkit: Practical techniques for building data warehouse and intelligent business systems.
4. Santos, V., & Belo, O. (2011). Slowly changing dimensions specification a relational algebra approach. In *Proceedings of the International Conference on Advances in Communication and Information Technology* (pp. 50–55).
5. Leonard, A., Mitchell, T., Masson, M., Moss, J., & Ufford, M. (2014). Slowly changing dimensions. In SQL *server integration services design patterns* (pp. 261–273). Apress.
6. Ma, K., & Yang, B. (2015). Introducing extreme data storage middleware of schema-free document stores using MapReduce. *International Journal of Ad Hoc and Ubiquitous Computing*, 274–284.
7. Dean, J., & Ghemawat, S. (2008). MapReduce: Simplified data processing on large clusters. *Communications of the ACM, 51*(1), 107–113.

Part II
Social Networking

Chapter 3
Intelligent Web Data Management of Social Question Answering

3.1 Introduction

3.1.1 Background

Currently, the growing popularity of social networking sites (SNS) among real-name Internet users changes the lifestyle and social behavior of human beings [1]. It is a platform to build social networks or social relations, such as interests and activities sharing. Most social network services are web-based and provide means for users to interact over the Internet. The main types of social networking services are those that contain category places, means to connect with friends, and a recommendation system linked to trust. Popular SNS services are Facebook, Google+, LinkedIn, Twitter, and WhatsApp.

Social question answering systems can benefit nearly every type of online business you can think of. It is a new system where users can find support online. With the development of Internet, more and more people turned towards the Internet to find support, and share their experience when they live difficult situation online.

3.1.2 Challenges and Contributions

Such sites make offline question answering online. There are examples of these applications, such as Facebook, help from senior brother (http://www.sxbbm.com) and ihelpoo (http://ihelpoo.com). However, there are some insufficiencies in the current applications. First, the success rate of current SNS-like online question answering system is relatively low. Users are used to saying their anxieties, needs and feelings, but would not like to offer answering actively. Second, online question answering is poor in time effectiveness. Generally, the urgent help published in the

© Springer International Publishing Switzerland 2016
K. Ma et al., *Intelligent Web Data Management: Software Architectures and Emerging Technologies*, Studies in Computational Intelligence 643,
DOI 10.1007/978-3-319-30192-1_3

system is not solved in time. For example, we need an umbrella owing to a storm coming on suddenly, trapped next to the teaching building. It is too late to post a help feed in the system. Third, we need to develop or purchase a browser-based system to deploy it in the Web servers with high cost and inconveniences.

To address these limitations, we design a new online social question answering architecture for multi-tenant mobile clouds. We propose a helper recommendation algorithm to find the helpers who are competent. In order to speed up the help process, we design a smart mobile client to post the help feeds based on location-based services (LBS) to find the helpers closest to you. Besides, we transform the legacy Web application into the multi-tenant architecture, which provides services that consumers rent. This pattern that clients only rent the service rather than purchasing the software is very convenient.

Considered to be part of the nomenclature of cloud Computing, software as a service (SaaS) is an emerging on-demand software delivery model [2, 3]. The feature of a multi-tenant mutual help architecture is data-intensive. The access of the data of this application is so frequent that read to write ratio is relatively high. Generally, large scale data are presented in this SaaS pattern. Although modern multi-tenant architecture has more advantages than legacy applications, this change has brought us with new challenges of online social question answering architecture for multi-tenant mobile clouds. First, we must consider how the multi-tenant big data are stored in the cloud. Second, we must think deeply how to improve the read/write efficiency of multi-tenant big data.

In the rest of this Chapter, we focus on the architecture of online social question answering system. The contributions of this chapter are several folds. First, we create a harmonious opportunity to help or be helped online. This can attract the users to rely on this system to solve offline issues. Second, this system architecture supports the multi-tenancy of the mobile cloud. The system we design can serve the tenants in the form of a service. Third, we propose a helper recommendation algorithm to assist online social mutual help. This process is composed of pre and post optimization. Some related help cases and helpers are recommended during the release of the help feed. Besides, we use the integration of push and pull. Fourth, we use key/tuple databases to store the help feeds to eliminate the read bottlenecks. It is also used to store the tenants' personalized information. Finally, the mobile architecture of our proposed mutual help system is loosely coupled using RESTful Service.

3.2 Related Work and Emerging Techniques

3.2.1 Social Question Answering

Question answering origins from offline human communication. Along with the Internet techniques, online social question answering is topic-oriented and procedural information delivered through software. Compared with offline question answering, online question answering is an effective attempt to present the right information to the right people at the right time in most effective and efficient form. Currently, most online question answering is designed to give assistance in the use of SNS-like system. Online question answering acts as an extension of offline question answering by providing quick answers to supplicants. There are many advantages of online question answering, such as instantaneity and expansibility. In this chapter, we take advantage of online social question answering to increase offline question answering. Before the release of question feed, we will seek some related solution by analyzing the historical knowledge base automatically. After the release of question feed, some referees are recommended.

Our online social question answering system provides a feed of the status of inserts, updates and deletes. There are two categories of driven approaches, which are shown in Fig. 3.1. The first approach is push pattern. In this solution, the help feed is pushed to the referees offering assistance. It will produce several records in case of one post. Although the feeds are well partitioned using shadings, the pushing amount of the feeds is very huge in an instant. Another approach is pull pattern. In this solution, only one feed and its relation are produced. Users can pull the feeds actively. Although the feeds are stored in the cache, the frequent query of the cache brings tremendous pressure to the application. The pull pattern is less efficient than the push pattern in the field of query performances. Another improvement of time-partitioned pull pattern. This pattern stores the feeds over a period of time in different partitions. The older question feeds are archived in the historical database. Frequent users will access the latest feeds. Compared with the current approaches, we benefit the help feed propagation from the integration of push and pull.

3.2.2   Multi-tenancy

Compared with multi-instance architectures, multi-tenancy refers to a principle in SaaS architecture where a single instance of the software runs on a server, serving

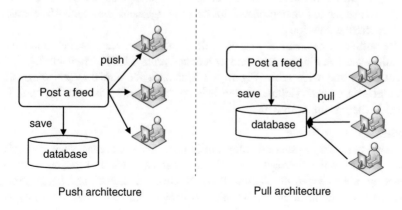

Push architecture   Pull architecture

Fig. 3.1 Push and pull architecture

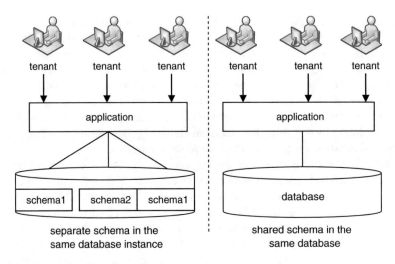

Fig. 3.2 Approaches for data management in a multi-tenant deployment

multiple tenants. With a multi-tenant architecture, a software application is designed to virtually partition its data and configuration, and each tenant works with a customized virtual application instance. There are many approaches for data management in a multi-tenant deployment, which are shown in Fig. 3.2 [4, 5]. The first approach is separate schema in the same database instance, housing multiple tenants in the same database but with different schemas. In this solution, each tenant has its own set of tables and views separately. The second approach is shared schema in the same database. In this solution, all the tenant data are mixed together with the same schema. For example, a given table can include records from multiple tenants stored in any order. A tenantID column associates every record with the appropriate tenant. This approach has the lowest hardware and backup costs. However, this approach may incur additional effort in the area of privacy and security to ensure that tenants can never access other tenants' data. After evaluating the pros and cons of two approaches for data management, we decided to pursue the schema sharing solution.

We outline some classical sharing multi-tenant data storage models and features in multi-tenant sharing system, and introduces private table, extension table, document store, and wide table. Figure 3.3 illustrates the structure and metadata of different data models. Column A and B is the customizing data, and column C, D and E is the personalized data.

Model 1: Private Table
The most basic way to support extensibility is to give each tenant their own private table. In this simple approach, what the query-transformation layer needs to do is renaming tables. Thus, this approach has stronger pertinence and better expansibility on the customization and isolation. However, only moderate consolidation is

Fig. 3.3 Classical
multi-tenant data models

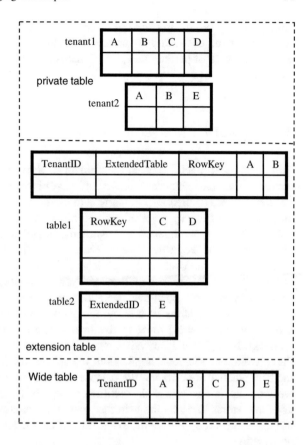

provided, since many tables are required. Some experts stated that relational
structures might be made more competitive in the case of over 50,000 tables.

Model 2: Extension Table
This approach is combined with splitting of the extension into a separate table. The
sharing data are stored in the public storage (basic tables), while the separate data
are stored in the extension table. Because multiple tenants may use the same
extension, extension tables as well as basic tables should be given an identity
column of a tenant ($tenantID$). However, this approach has not yet resolved the
expansion problem. It still needs a mechanism to map the basic and extended tables
into one logical table.

Model 3: Wide Table
Wide table is usually highly sparsely populated so that most data can be fit into a
line or a record. Using this solution, queries are composed on only a small subset of
the attributes. However, this model will produce the schema null problem, since it
has too many null values. Moreover, it cannot provide dynamic customization
capabilities since the reserved column is still limited in numbers. Indexing on a

table generally brings up extra high storage and update costs. When the data set is sparse, the extra cost can be overcome by using a sparse index. A sparse index is one special kind of partial index, which maps only the non-NULL values to object identifiers.

3.2.3   NoSQL Storage

Currently, most practitioners in the domain of data management systems are discovering that one size does not fit all. Despite the disadvantages of using a general-purpose RDBMS in comparison to more specific solutions, we expect that a large amount of legacy applications will remain in the years ahead. Motivated by horizontal scaling and finer control over the availability, NoSQL databases are often highly optimized key/value stores intended for simple retrieval and appending operations, with the goal being significant performance benefits in terms of latency and throughput [6]. As a new type of NoSQL, key/tuple stores are intended for simple retrieval and appending operations, with the goal being significant performance benefits in terms of latency and throughput. This is advantageous in the significant and growing industry use in big data and real-time web applications. Although NoSQL has many advantages, it is an enormous project to migrate from the conventional relational applications to support NoSQL transparently. Therefore, the best practice is achieving the balance of RDBMS and NoSQL. In our chapter, we adopt key/tuple store as the auxiliary storage of RDBMS. On one hand, we use key/tuple stores to cache the help feeds and their relations. On the other hand, we use key/tuple stores to save the extended vertical metadata of tenants' personalized information.

3.2.4   RESTful Web Service

Recently, REpresentational State Transfer (REST) [7] appeared as an emerging abstract architectural style. This emerging technology involves providing a new abstraction for publishing information and giving remote access to applications. This abstraction provides the foundation for so-called RESTful Web services. It can be directly mapped to the interaction primitives, making it easy to integrate with the third-party systems.

In the RESTful solution, the set of operations are divided into four methods (PUT, GET, POST, and DELETE respectively) once a resource has been identified. These create, read, update, and delete (CRUD) operations apply to all resources with similar semantics, even if in some cases not every operation can be meaningful. Each of these methods is invoked on a resource using synchronous or asynchronous HTTP request–response interaction.

3.3 Requirements

We have the following requirements while designing online social question answering architecture for multi-tenant mobile clouds.

- **High success ratio**: Except attracting more tenants and users to rent and use our system, we want to ensure that the strategy of the system has a high success rate.
- **Help feed propagation efficiency**: We want to improve the performance of feed propagation, spreading the published help feeds to the helpers.
- **Multi-tenancy**: Since our system provides services with the tenants, the architecture and implementation of multi-tenancy is the prerequisites.
- **Tenant customizing**: Besides the multi-tenancy, we want to provide the tenant customizing of data and applications.
- **Open internetwork**: Since the social mutual help system we design provides the API for different clients, we want to provide RESTful Web Service API to integrate with different heterogeneous systems.

3.4 Architecture

In this section, we talk about the architecture of this online social question answering system for multi-tenant mobile clouds. The proposed helper recommendation algorithm, help feed propagation method, and some key techniques are discussed in detail.

3.4.1 Helper Recommendation Algorithm

Online social question answering is the core business of this system. We attempt to optimize the traditional online social question answering process from two perspectives, which is different from the current online help products.

Perspective 1: Pre-process of recommendation
Before the supplicant posts a question feed, our system will match the content of this question feed with the knowledge base, which is composed of the successful question feeds. To assure the question answering is with a high level accurate and success, we classify every successful question answering knowledge with tags and categories. We use Levenshtein distance algorithm [8] to measure the difference between the descriptions of this question content with the knowledge base. The objective is to find matches for short strings in many longer descriptions of the help. The match algorithm is shown as follows.

The Levenshtein distance between two strings a and b is given by $lev_{a,b}(i, j)$ where $0 < i < length(a)$, $0 < j < length(b)$, and

$$
lev_{a,b}(i, j) = \begin{cases} max(i, j) & \text{if } min(i, j) = 0 \\ min \begin{cases} lev_{a,b}(i-1, j) + 1 \\ lev_{a,b}(i, j-1) + 1 \\ lev_{a,b}(i-1, j-1) + [a_i \neq b_i] \end{cases} & \text{otherwise} \end{cases}
$$

The first element in the minimum corresponds to deletion (from a to b), the second to insertion and the third to match or mismatch, depending on whether the respective symbols are the same. The similarity of these two help feeds is similarity = $1-$ $lev_{a,b}$(length(a), length(b))/max(length(a), length(b)). We have made some tests to conclude that the threshold of the similarity is 0.6. After the match with the knowledge base, the help cases with over 0.6 similarity are provided with the supplicant as a reference. The system we design will remind the supplicants that they can stop this process if the system matches a reference solution automatically. Compared with current similar products, this strategy sometimes overcome the difficulty in advance.

Perspective 2: Post-process of recommendation
After the supplicant posts a question feed, our system will begin to work on data mining and machine learning of the user information and historical help behavior. We believe that the users with similar habits and behaviors are more likely to offer social answering. Therefore, we propose a helper recommendation algorithm based on user affinity. After that, the recommended helpers with high affinity are given to the supplicant to accelerate the answering process.

The user affinity reflects the relationship of the users. We define the user affinity with the integration of basic affinity, interactive affinity, and similarity affinity. The basic affinity is defined as $B(X, Y) = \Sigma(w_i \times B_i)$, where w_i is the coefficient of the aspect B_i. For example, mutual friend and help frequentness are the influence factors. We use the social networking graph $G = (N, E, W)$ to define interactive affinity, which reflects the behaviors of users. N is the set of user nodes, E is the set of directed behavior edges, and W is the weight of the edges. We consider the minimum weight as the interactive affinity, denoted as $E(X, Y) = min(W(X, Y), W(Y, X))$. The user nodes without the direct link may also has some affinity. We call this similarity affinity, denoted as $S(X, Y) = \Sigma(w_i \times S_i)$. For example, user habits and personal information are the influence factors. We define the user intimacy depending on the summation of the above factors, denoted as $F(X, Y) = \alpha \times B(X, Y) + \beta \times E(X, Y) + (1 - \alpha - \beta) \times S(X, Y)$, where α $(0 < \alpha < 1)$ and β $(0 < \beta < 1)$ are the important factors.

We use the user affinity to rank the ratings of the potential helpers. The helper with the largest user affinity is most likely to help the supplicant. Next, we sort the recommended helpers by the user affinity. Those in the front are recommended to the supplicants. Compared with current similar products, this strategy improves the success rate to some extent.

3.4.2 Help Feed Propagation Method

With end-user license agreement, our smart mobile with Global Positioning System (GPS) and assisted GPS (AGPS) module can submit the user's location computed by location-based services (LBS) automatically. This will help the supplicants to find the closest helper. In case of some environment without GPS, the location can be entered manually by the user.

We use the integration of push and pull to assist help feed propagation [9]. The architecture is shown in Fig. 3.4. After the post of a feed, the feed is put in the post help queue. We use the key/tuple stores as the cache of help feeds to accelerate the read/write efficiency. Next, we propose an optimization scheme to the feed propagation. We classify the feed propagation into two categories. This integration strategy greatly reduced the amount of needless pushing and pulling, decreasing the performance loss. Besides, we use the key/tuple stores as the cache of posting relations (feed, from_relation and to_relation).

- In order to reduce the amount of pushing data, the system will push the help feeds to the potential helpers actively.
- The inactive potential helpers will pull the help feeds actively when they login this system next time.

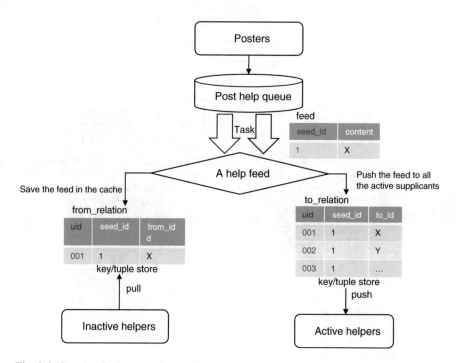

Fig. 3.4 Question feed propagation architecture

3.4.3 *Multi-tenancy*

The typical multi-tenant architecture of our system is shown in Fig. 3.5. The architecture is composed of three points of view in order to reduce the complexity: application layer, transformation layer, and database layer.

- From the top is application layer, where tenants rent our online mutual help system.
- At the bottom is database layer, where we use shared key/tuple stores as the storage.
- In the middle is transformation layer. On one hand, we extract the tenantID from the Web context. On the other hand, the search criteria and change update (insert/update/delete) of online mutual help system is intercepted by the interceptor component. For the query of the application, tenantID is connected with the search criteria to formulate the new criteria. For the change update, tenantID is added as the new criteria to the update.

As shown in Fig. 3.6, our multi-tenant architecture is logically isolated for each tenant with higher demand on security. On one hand, we provide a logical view for each tenant to read. All the data that do not belong to this tenant are filtered by

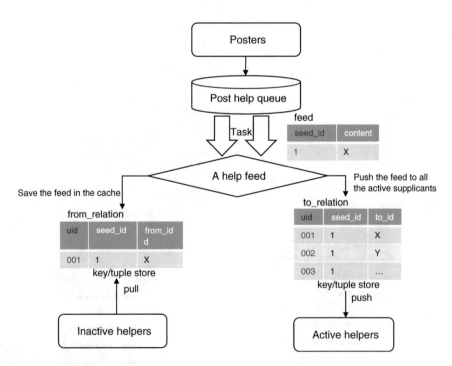

Fig. 3.5 Multi-tenancy architecture

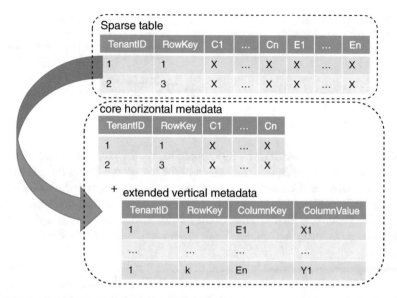

Fig. 3.6 Logical isolation of multi-tenancy architecture

tenantID automatically. On the other hand, we provide a transformation engine to change all the writes of a tenant transparently. This engine adds a tenantID attribute to one record or document.

3.4.4 Data Customizing of Tenants

There are some features of multi-tenant data due to sharing personalized customization. First, there will produce the schema null issues when the tenants do not customize some columns. This is a key issue of the data storage of the multi-tenants. Second, the multi-tenant data are left-intensive and right-sparse, since the tenants consume the data from left to right.

Compared with sparse table, we propose an integrated data model of core horizontal and extended vertical metadata (hereinafter integrated data model), which can solve the sparse and schema null issues at the same time. That is to say that we extract the personalized data from the sparse table, and then describe it using vertical metadata (hereinafter extended vertical metadata). Figure 3.7 shows the transformation process. As depicted in Fig. 4.7, extended vertical metadata efficiently reduces the waste of data resources. Each row in the extended vertical metadata is a key/value pair, which is used to store the customization of tenants to fulfill the requirements of different tenants. In the case that the customization of tenants is the same, the extended vertical metadata can be omitted.

Fig. 3.7 Transformation
from sparse table to integrated
data model

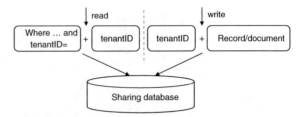

We use core = (tenantID, rowKey, columns) as the core horizontal metadata, where tenantID is the unique identification of the tenant, rowKey is the primary key of the wide table, and columns is a set of columns. We use extended = (tenantID, rowKey, columnKey, columnValue) as the extended vertical metadata, where tenantID is the unique identification of the tenant, rowKey is the primary key of the wide table, columnKey and columnValue are the key/value of the extended vertical metadata.

Although the multi-tenant integrated data model reduces the schema null effectively, it increases the computational complexity. For the purpose of evaluating whether the data are stored in the core horizontal metadata or the extended vertical metadata, we give the evaluation function to determine whether the extended vertical metadata are used. Given F_k as the evaluation function of the k_{th} column, where μ_k is the proportion that different tenants customize the kth column, α_k is the access number of the kth column, β_k is the access number of the tables containing the kth column, and γ_k is the service factor that the tenant serves the kth column. We can conclude that $F_k = \mu_k * \alpha_k/\beta_k + (1 - \mu_k) * \gamma_k$, where $\mu_k = (\sum(k \geq C_i)? 1 : 0)/n$. The larger F_k is, the less appropriate the kth column is in the extended vertical metadata. The data intensity of multi-tenant integrated data model is much less than sparse table.

In the online mutual help system, we use the key/tuple stores as the cache of help feeds and their relations. Feed, from_relation and to_relation are saved in the key/tuple stores.

3.4.5 RESTful Web Service API

We design the core business to implement the different clients (desktop client, mobile client, and Web user interface) using RESTful Service API. This allows for easy integration with existing third-party system. The communication uses the form in API with the JavaScript Object Notation (JSON) format to communicate with each other. Table 3.1 summarizes its provided functionalities.

Table 3.1 Online mutual help service resources

Resources	URL	Parameters	Description
List of knowledge base	GET: /knowledge/	tenantID=?&uid=? &content=?	Find the matched help feeds from the knowledge base
List of users	GET: /users/	tenantID=?&uid=? &content=?	Find the matched user who can help him
List of help feeds	POST: /feeds/	tenantID=?&uid=? &content=?	Post a help feed
List of to_relations	POST: /to_relations/	tenantID=?&uid=? &sid=?	Push a help feed
List of from_relations	POST: /from_relations/	tenantID=?&uid=? &sid=?	Pull a help feed

3.5 Evaluation

We conducted the experiments on our physical machine: 4 core 2.80 GHz Intel Core (TM) machines with 8 GB RAM, 128 GB SSD and 100 Mbps Ethernet. The system was configured with a Windows Server 2012 × 64. We adopt Redis 2.8.5 as the key/tuple store, and MySQL 5.6.16 as the relational database. We deployed the prototype system with the link http://aite5.com. The screenshot of the developed online social mutual help architecture is shown in Fig. 3.8.

Figure 3.9 shows an example of interactive process of online social question answering. First, the supplicant in trouble posts a plea feed for help "I need an umbrella downstairs" using her smart mobile. With the user's explicit consent, we collect the location information to propagate this plea to the potential helpers. Our helper recommendation and propagation algorithms are used to assist and complete this help successfully. A nearby user using this system might appear with an umbrella to adopt this plea. Some backend processing that does not appear in the user interface are the pre and post recommendation. First, the system will discover the successful cases in the knowledge base. In this example, the pre-process fails since this is not a knowledgeable help. This system will rank all the potential helpers by user affinity, and push this help to the helpers who are most likely to complete this task.

3.5.1 High Success Ratio

First, we have made an experiment to test the success ratio of the mutual help process. We obtain the data in the past six months. For the first three months, we

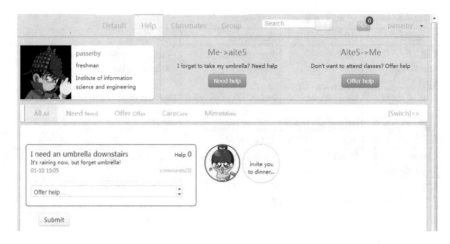

Fig. 3.8 User interface of online social mutual help (Due to privacy issues, we have replaced the photos with cartoon drawings)

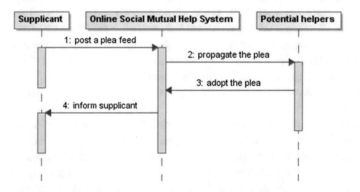

Fig. 3.9 An example of interactive process of online social mutual help

deploy this social mutual help system without any helper recommendation. For the last three months, we deploy this social mutual help system with helper recommendation optimization. We record the amount of published help feeds to calculate the success ratio. As illustrated in Table 3.2, the helper recommendation algorithm

Table 3.2 Help process in the past six months

Version	Name	Period of time	Amount	Success amount	Success ratio (%)
1.0	Without helper recommendation	First 3 months	73	52	71.5
2.1	With helper recommendation	Last 3 months	95	78	82.1

indeed improves the success ratio of help feeds. Besides, we attract some more users to involve in the mutual help process (growth rate of help feeds 26.7 %).

3.5.2 Propagation Time

After the helper recommendation, some potential helpers are calculated. We have made an experiment to measure the propagation time of help feeds using different solutions. Considering that the amount of potential helpers is 10, 100, 1,000, 10,000 and 100,000 respectively. We measure the propagation time (the period of time between the release and receipt of the help feed) for further analysis. The percentage of active users in our system is about 48.7 %. As depicted in Fig. 3.10, the propagation time of pull solution is the shortest, since this solution only saves one relation no matter how many helpers the supplicant has. The propagation time of our integration solution is shorter than push solution. Although the pull solution seems superior to our solution, users cannot get the real-time notification on the help feed in fact. They will see this plea when they login this system next time.

3.5.3 Propagation Space

Next, we have made an experiment to measure the propagation space of help feeds using different solutions. Considering that the amount of potential helpers is 10,

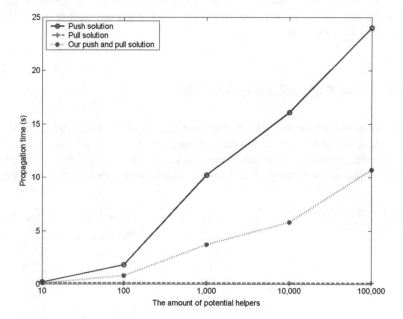

Fig. 3.10 Comparison of propagation time

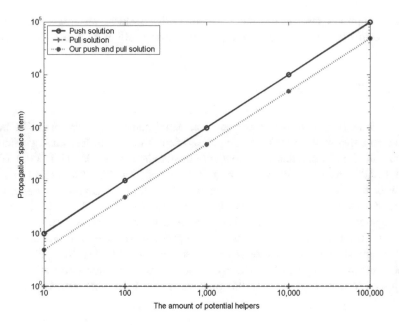

Fig. 3.11 Comparison of propagation space

100, 1,000, 10,000 and 100,000 respectively. The percentage of active users in our system is about 48.7 %. As depicted in Fig. 3.11, the propagation space of pull solution is the smallest, since this solution only saves one relation no matter how many helpers the supplicant has. The propagation space of our integration solution is smaller than push solution. Due to the delay users get the notification, the pull solution is not suggested.

3.5.4   Data Customizing of Tenants

We have made an experiment to measure the sparsity of the integrated multi-tenant data model. Consider tenants consume the columns from left to right. In order to quantize the sparsity, we suppose the amount of different tenants is the same. Table 3.3 shows the sparsity of each tenant. The average sparsity of sparse table is

Table 3.3 Data customizing of different tenants

Tenant	Common column	Customizing column	Sparsity (%)
1	10	2	83.3
2	10	4	71.4
3	10	3	76.9
4	10	5	66.7
5	10	7	58.8

70.4 %. With the help of integrated multi-tenant data model we propose, we store the common columns in the core horizontal metadata, and the customizing columns in the extended vertical metadata. The sparsity of this model is nearly 100 %. The experiment results show that the sparsity of integrated multi-tenant data model is larger than the sparse table. In order to improve the performance of integrated multi-tenant data model, we create a view using PIVOT operation. This view is logically the same with the sparse table.

3.6 Discussions

3.6.1 Preprocessing of Successful Knowledge Base

We take some measures to optimize the successful knowledge base. First, we merge the answers of the similar question. Second, we merge the questions of the similar answer.

3.6.2 Expert Discovery

We sort out the helpers by the frequency of answering providers in one topic. If the helper is willing to help others, it is recommended as an expert in this field.

3.7 Conclusions

In this chapter, we design an online social question answering architecture for multi-tenant mobile clouds. We proposed a helper recommendation algorithm to accelerate the help process. Then we presented the integration solution of push and pull to propagate the feeds to the potential helpers. Finally we provided some RESTful Web service APIs for the integration with the third-party systems. The experimental results illustrate that this architecture improves the success rate and performance successfully and provides the support of data customizing of tenants to increase the sparsity.

References

1. Moreno, M. A., Jelenchick, L. A., Egan, K. G., Cox, E., Young, H., Gannon, K. E., & Becker, T. (2011). Feeling bad on Facebook: Depression disclosures by college students on a social networking site. *Depression and Anxiety, 28*(6), 447–455.
2. Armbrust, M., Fox, A., Griffith, R., Joseph, A. D., Katz, R., Konwinski, A., et al. (2010). A view of cloud computing. *Communications of the ACM, 53*(4), 50–58.

3. Allen, B., Bresnahan, J., Childers, L., Foster, I., Kandaswamy, G., Kettimuthu, R., et al. (2012). Software as a service for data scientists. *Communications of the ACM, 55*(2), 81–88.
4. Ma, K., Yang, B., & Abraham, A. (2012). A template-based model transformation approach for deriving multi-tenant saas applications. *Acta Polytechnica Hungarica, 9*(2), 25–41.
5. Ma, K., & Yang, B. (2014). Multiple wide tables with vertical scalability in multitenant sensor cloud systems. *International Journal of Distributed Sensor Networks, 2014*.
6. Cattell, R. (2011). Scalable SQL and NoSQL data stores. *ACM SIGMOD Record, 39*(4), 12–27.
7. Allamaraju, S. (2010). *Restful web services cookbook: solutions for improving scalability and simplicity*. O'Reilly Media, Inc.
8. Levenshtein, V. I. (1966, February). Binary codes capable of correcting deletions, insertions, and reversals. In *Soviet physics doklady* (Vol. 10, No. 8, pp. 707–710).
9. Ma, K., & Tang, Z. (2014). An online social mutual help architecture for multi-tenant mobile Clouds. *International Journal of Intelligent Information and Database Systems, 8*(4), 359–374.

Chapter 4
Intelligent Web Data Management of Content Syndication and Recommendation

4.1 Introduction

4.1.1 Background

As one form of Web 2.0 technology, really simple syndication (RSS), known as rich site summary, uses a family of standard web feeds to publish frequently updated information, such as blog articles and news headlines [1, 2]. An RSS document (called feed or channel), including full or summarized text and metadata of information, enables publishers to syndicate data automatically. In this way, subscribing to an RSS source removes the need for the user to manually check the web site for new content.

Currently, there are several ways to obtain information through the Internet. The first popular way is to find a search engine, such as Google, Yahoo and Bing. The second way is to browse the Web portal of news websites, such as the wall street journal and China daily. The third way is the information transfer using social network sites, such as Facebook, Twitter and MySpace. In the era of mobile computing, the information is pushed to the user by smart devices, such as mobile, tablet, and laptop.

There are some drawbacks of current methods. The searching results of the searching engine are inundated with advertisements, advertorials, and fake information. It is difficult to pick out the useful and valuable message from the miscellaneous information. Although amount of people like to read news chapter and browsing news websites, it is difficult to guarantee the quality of the content. Regarding the social networking website, the information is pushed to the user passively without considering its value. The same problem also exists in the context of the mobile applications of smart devices. Therefore, the push of the information need a filtering mechanism before sending to the users. That is to say that the requirements on the customizing is necessary.

© Springer International Publishing Switzerland 2016   65
K. Ma et al., *Intelligent Web Data Management: Software Architectures and Emerging Technologies*, Studies in Computational Intelligence 643, DOI 10.1007/978-3-319-30192-1_4

4.1.2   Challenges and Contributions

Although RSS aggregator is commanded to automatically download the new data, there are still some challenges of this approach. First, RSS feed is organized within the form of formatted XML item. Therefore, the search becomes a bottleneck issue when encountering massive data. Furthermore, current RSS products have weak ability to support cross-source search. Second, RSS is a hard-to-replicate source of information. Some RSS products are designed to enable asynchronous synchronization of new and changed items amongst a variety of data sources. For example, some products implement timestamp-based incremental synchronization with some challenges. Therefore, a set of algorithms followed by all endpoints to create, update, merge, and conflict resolve all items are a pressing need to handle the synchronization issue. Third, although the content syndication has replaced traditional search through various categories of a website in order to find interesting articles successfully, users find it difficult to obtain the useful information along with an increasing number of subscribed RSS feeds. An intelligent recommendation algorithm is designed to filter and sort the massive feeds before pushing to users. However, current RSS products lack of automatic collection and classification. To address these limitations, we propose a new architecture of content syndication and recommendation.

In the rest of this chapter, we focus on the architecture of content syndication and recommendation system. The contributions of this chapter are into several folds. First, we create a user-friendly content syndication and recommendation architecture. This can actively push valuable information users are interested in instead of traditional pulling technology. This push technology enables that the feeds are filtered and sorted by the interests and hobbies of users. Second, we design an RSS source listener to capture the incremental feed changes of multiple RSS sources in a very short time. Third, we design schema-free documents to persist the feeds to speed up the search efficiency. Finally, we design OAuth2-authorization RESTful feed sharing APIs to be integrated with the third-party system.

4.2   Related Work and Emerging Techniques

4.2.1   RSS Specification

RSS is the abbreviation Really Simple Syndication, which uses a family of standard web feed formats to publish frequently updated information: blog entries, news headlines, audio, and video. An RSS feed (often called channel) includes full or summarized text, and its metadata with the compatible XML file format. RSS feeds enable publishers to syndicate data automatically. Furthermore, RSS feeds also benefit users who want to receive timely updates from favorite websites or to aggregate data from many sites. There are several different versions of RSS, falling

Table 4.1 RSS' specification and version

Specification	Version
Really simple syndication	RSS: 0.91, 0.92, 0.93, 0.94, 2.0
Rich site summary	RSS: 0.91, 1.0

into two major branches: RDF and 2.X, where RDF is also called RSS 1.X. Currently, the most widely used standard is RSS 2.X [1]. Table 4.1 shows all RSS versions.

4.2.2   RSS Products

Currently, there are some open-source and commercial RSS products (also called reader or aggregator).

Google Reader was a web-based aggregator, capable of reading RSS feeds online or offline. Due to declined usage, Google powered down this product [3].

Feedly [4] is another alternative RSS reader, which is a better way to organize, read and share the content of RSS sources. It weaves the content from the RSS feeds of your favorite websites into a fun magazine-like experience and provides seamless integration with social networks.

Reeder [5] is an RSS reader and client for Feedly, which caches articles and images from your feed. However, the cache will encounter the bottleneck issue in the case of massive feeds. Recently, some emerging RSS readers have supported the integration with social networking application researching on the recommendation algorithm. But the storage of this method to support millions of feeds is not clear.

However, there are some deficiencies of current RSS products. First, since it is not a good method to search the feeds in the XML file, most of current RSS products adopt the relational database as the storage engine. This will improve the query performance in the case of small data quantity. When the amount of data is large, search in a traditional relational database encountered the bottleneck. Although some RSS products enable cache to optimize the storage, the effect is not obvious in the case of big data. The query of feeds is always in the single RSS source, while it is not clear to support the cross-source search. The second limitation is lack of pre-processing before presenting to the users. Once a user subscribes to several RSS sources, it means all the untreated information is pushed to the user. This information consists of advertisements, advertorials and fake information. To improve user experience, the information should be sorted and filtered by the interest of users first.

4.2.3   Feed Synchronization

Currently, there are some patents on feed synchronization. The first is timestamp-based feed synchronization. The RSS source listener intercepts RSS feed changes to extract the timestamp, and determine whether this feed is updated from the last synchronization timestamp to this current timestamp. However, this method takes much spaces to store massive time records for the subsequent synchronization. Although this method can shorten the synchronization time, it records too much timestamps in the case of frequent updates. Another is clock-based feed synchronization. The publisher creates the feed by including a media content associated with therewith. The first virtual clock value is provided to the subscriber to modify the first virtual clock value when the subscriber modifies the media content associate with the web syndication item. However, this additional attribute of RSS feed does conform to RSS specification. On the contrast, we propose the synchronization method based on the unique key (composition of guid and pubDate of a feed) of the feed.

4.2.4   RSS Recommendation

Currently, researchers have focused on the recommendation algorithm on social networking. Since RSS is content-centered rather than social networking system, research on RSS recommendation is scarce. InterSynd is simple Web client to recommend new RSS feeds to users based on what their neighbors have subscribed to. FeedMe [6] is another system that filters alerts with a combination of collaborative filtering technique and naive Bayes classifier. Moreover, some experts firstly used probabilistic latent semantic analysis (PLSA) to discovery the topics of blog posts, then adopted Naive Bayesian algorithm to classify the blog posts. The goal is to reduce the noise caused by unwanted interruptions. In contrast, our goal is to recommend feeds to users and we focus on matching the feed contents with users' interest.

4.3   Requirements

We have the following requirements while designing a content syndication subscription architecture.

- **Cross-source search**: We want to ensure that the system has the ability to support search on cross RSS sources.
- **High efficient search**: Except supporting heterogeneous RSS sources, we want to improve the performance of feed search, but not limited to cache technology.
- **Feed recommendation**: We want to push the feeds most concerned to the users, and other useless information is discarded.
- **Open internetwork**: Since the content syndication and recommendation system we design provides the API for different clients, we want to provide an OAuth2-authorization [7] RESTful feed sharing solution to be integrated with different heterogeneous systems.

4.4   Architecture

In this Section, we talk about the architecture of content syndication and recommendation system, which is shown in Fig. 4.1. The source listener, feed search, feed recommendation, and OAuth2-authorization RESTful feed sharing APIs and some key techniques are discussed in detail. First, multiple RSS sources generate a set of RSS feeds, and then source listener captures the feed updates to propagate them to the schema-free document stores. Second, we utilize TF-IDF algorithm [8] to conclude the keywords. They are matched with the users' interest to re-rank all the recommended feeds. Finally, the results are pushed to the users.

4.4.1   Source Listener

Source listener is an important part of this system, which is responsible for propagating feed changes to the schema-free feed repository incrementally. We propose an optimized source listener to capture the increment. Initially, our source listener reads the latest information of an RSS source, and then store them in the schema-free repository for the first time. Next, this source listener only intercepts the changes to complete an incremental synchronization. In RSS 2.0 specification [1], there is an optional "guid" attribute to indicate the unique string of an RSS feed. If the feed of RSS source has this attribute, we take this as the unique key. If the feed of RSS source does not have this attribute, we take the composite source identity and timestamp as the unique key. After the RSS listener intercepts a new item, we make the incremental changes on the target schema-free repository according to this unique key. We call this source incrementality. We take polling monitoring method to intercept the frequent changes of multiple different RSS sources. We adjust the interval time on the basis of the update frequency.

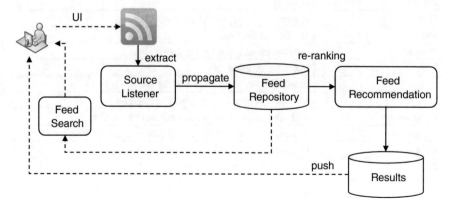

Fig. 4.1 Architecture

4.4.2   Feed Search

The intercepted feeds from RSS sources are stored in the form of schema-free documents. The RSS feed has two required attributes (guid, title and description), and several optional attributes (author, pubDate, link, et al.). With the rapid development of Web 2.0 sites, traditional relational databases in dealing with Web 2.0 sites encounter the bottleneck problem in the context of search. Therefore, we adopt schema-free documents to store all the updated feeds. A JSON-like document with dynamic schema is designed to store an RSS feed. The structure of this schema-free document is shown in Table 4.2. On one hand, separate index entry is created for each keyword (title, pubDate or category). On the other hand, full-text index is created on the large field (description) of a collection. Due to the difference of feed structure, it is difficult to implement the search on cross sources directly. Therefore, the captured feeds are stored in schema-free documents to keep a unified structure. The search issue might converted into the query on the target schema-free documents.

4.4.3   Feed Recommendation

Feed recommendation adopts a content-based recommendation technique, by mining the keywords of the contents, and matching them with the interests of subscribers'. The process by which this architecture generates a set of ranked RSS feeds is presented in detail in Fig. 4.2.

Since there is no keyword in the feed structure, we adopt classical text-based TF*IDF algorithm to conclude the keyword from the feeds automatically. The

Table 4.2 Structure of an RSS feed

Field	Description
Guid	A string that uniquely identifies the item
Title	The title of the item
Description	Description and The item synopsis
Link	The URL of the item
Author	Email address of the author of the item
Category	Includes the item in one or more categories
Comments	URL of a page for comments relating to the item
Enclosure	Describes a media object that is attached to the item
Pubdate	Indicates when the item was published
Source	The RSS channel that the item came from

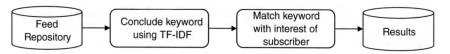

Fig. 4.2 Process of feed recommendation

keyword extraction is conducted exploiting the TF*IDF weight of the term. It is calculated according to the formula: TF*IDF(term) = TF(term)*log(1 + N/DF(term)), where TF(term) is the frequency of a term in the given feed, N is the total number of feeds in the collection, DF(term) is the number of feeds that contain this term.

Next, we rank RSS feeds to the subscriber. First, the keywords of each subscriber are compared the calculated keyword in the last step. We adopt Levenshtein distance algorithm to measure the difference of the keywords. The objective is to find matches for short strings in many longer descriptions of the feed. Finally, the ranked feeds are sorted out by the Levenshtein distance to push to the users.

4.4.4   OAuth2-Authorization RESTful Feed Sharing APIs

We provide an OAuth2-authorization RESTful feed sharing APIs for the third-party system to access RSS feeds on behalf of a subscriber. Using this integration solution, users simply issue access tokens rather than the password to access their subscribed feeds for sharing. The RESTful Service is designed to obtain the subscribed feeds of a subscriber. This allows for easy integration with existing third-party system. The communication uses the form in API with the JavaScript Object Notation (JSON) format to communicate with each other. Table 4.3 summarizes this interface.

As depicted in Fig. 4.3, the sequence starts (1) with a user requesting some service from the third-party system, and ends with the sharing feeds from our RESTful feed sharing service. The third-party system responds by (2) redirecting the end user's browser to a URL of OAuth2-authorization server. After verifying the identity of the user, the third-party system request access token from the OAuth2-authorization server. Then the third-party system can interact with the resource server to exchange the sharing feeds.

Table 4.3 Feed authorization API

Resources	URL	Parameters	Description
List of sharing feeds	listFeeds:/feeds/	uid=?&token=?	Find the sharing RSS feeds

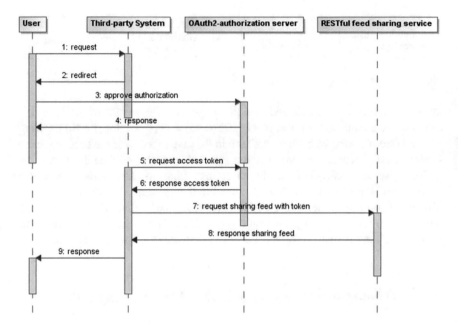

Fig. 4.3 Sequence diagram of interaction with our OAuth2-authorization RESTful feed sharing service

4.5   Evaluation

We conducted a set of experiments to evaluate the efficiency of the proposed content syndication and recommendation architecture on our physical machine: 4 core 2.80 GHz Intel Core (TM) machines with 8 GB RAM, 128G SSD and 100 Mbps Ethernet. The system was configured with a Windows Server 2012 x64. We adopt MongoDB 2.4.9 x64 as the document store. After a description of the experimental setup, we illustrate low latency of search, incremental synchronization, and user experience. We have deployed the prototype system with the link http://rsscube.duapp.com. The screenshot of the developed content syndication and recommendation system is shown in Fig. 4.4.

4.5.1   Low Latency of Search

The first experiment is to measure the search time with different amounts of feeds. Figure 4.5 shows the average search time of RDBMS and document store. We issue a query "obtain the feeds in the past three months by the keywords" for many times. With less than 50,000 feeds, the search time of document store is close to RDBMS.

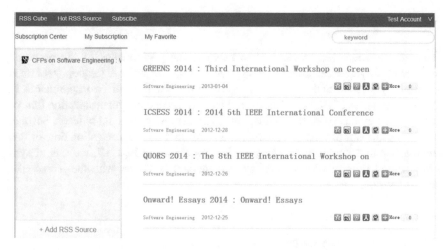

Fig. 4.4 Screenshot of our content syndication and recommendation system

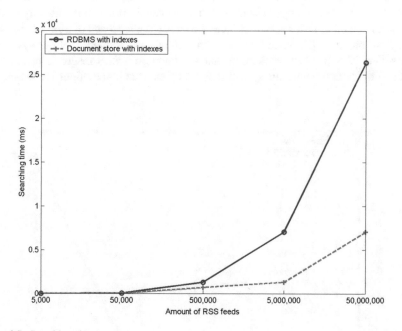

Fig. 4.5 Searching time

Along with the increasing number of feeds, the average search time of document store rises more sluggishly than the RDBMS solution. That is because document stores have powerful performance on the query.

4.5.2   Incremental Synchronization

Two experiments were conducted to evaluate the performance of the proposed incremental synchronization of RSS feed. First, we fix the frequency of feed updates by 1000/s, and measure the synchronization time of two approaches in different amounts of feeds. As shown in Fig. 4.6, the rate of synchronization time of our unique key-base method is slower with the increase of amount of feeds. Second, we fix the amount of feed by 500,000, and measure the synchronization time of two approaches in different update frequencies. As shown in Fig. 4.7, the rate of synchronization time of our unique key-base method is slower with the increase of frequency rate.

4.5.3   User Experience

We asked them to rank the two systems: one is a famous traditional RSS reader (Due to licensing and copyright restrictions, we cannot disclose the identity of this product); the other is our system. Figure 4.8 shows the feedback of user satisfaction. We have received 127 valid feedback. 66.9 % of users thought that our system is friendlier, 22.8 % of users thought that the traditional RSS reader is better, and 10.3 % of them thought our system may be better after some minor rectifications.

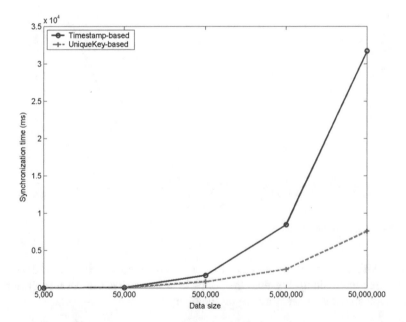

Fig. 4.6 Synchronization time with the increase of amount of feeds

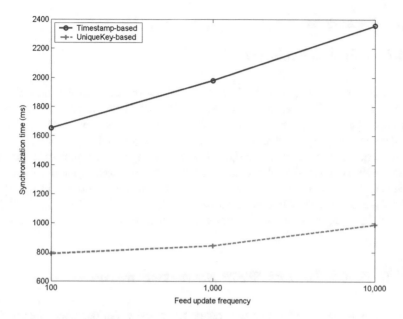

Fig. 4.7 Synchronization time with the increase of frequency rate

Fig. 4.8 User satisfaction

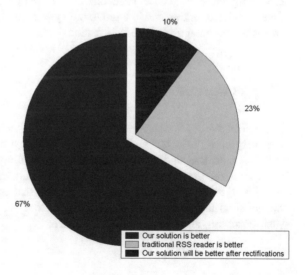

The approach of our solution has some advantages to user experiences. First, this active push mechanism simplifies the process of finding useful information from massive RSS feeds. Second, only the information of interest is pushed to the user. Moreover, our system re-ranks the sequence of the feeds using proposed recommendation algorithm.

4.6   Discussions

4.6.1   RSS Feeds Storage

With the rapid development of Web 2.0 sites, traditional relational databases in dealing with Web 2.0 sites, especially large scale social networking sites, encounter the bottleneck problem in the context of query. Due to its own characteristics of NoSQL, we adopt this database to store all the updated feeds.

We choose the popular NoSQL database MongoDB as our data storage. Mon-goDB is a cross-platform document-oriented database system. MongoDB eschews the traditional table-based relational database structure in favor of JSON-like documents with dynamic schemas (MongoDB calls the format BSON), making the integration of data in certain types of applications easier and faster.

4.6.2   Interaction with Social Networking Website

In order to assess the quality of the content more easily. On one hand, we provide an independent review and rating system. On the other hand, we blended social networking websites, such as Facebook and Google plus. Users can use the account of the social networking site to login to our system through OAuth 2.0, so that users can easily comment and share information.

4.7   Conclusions

In an era of Web 2.0 and big data, how to access the information efficiently and effectively becomes a hot topic. Therefore, we have proposed an architecture of content syndication and recommendation, which is composed of source listener, feed search, feed recommendation, and OAuth2-authorization RESTful feed sharing APIs. Finally, experimental results illustrate that content syndication and recommendation architecture has some features: low latency of search, incremental synchronization, and friendly user experience.

References

1. Board, R. A. (2007). RSS 2.0 speciation (version 2.0.10).
2. Ovadia, S. (2012). Staying informed with really simple syndication (RSS). *Behavioral and Social Sciences Librarian, 31*(3–4), 179–183.
3. Benzinger, B. (2009). *Google Reader Reviewed*. Retrived from October 7, 2005.
4. Feedly. http://www.feedly.com.

5. Reeder. http://reederapp.com.
6. Sen, S., Geyer, W., Muller, M., Moore, M., Brownholtz, B., Wilcox, E., & Millen, D. R. (2006). FeedMe: a collaborative alert filtering system. In *Proceedings of the 2006 20th Anniversary Conference on Computer Supported Cooperative Work* (pp. 89–98). ACM.
7. Hardt, D. (2012). The OAuth 2.0 authorization framework.
8. Zhang, W., Yoshida, T., & Tang, X. (2011). A comparative study of TF*IDF, LSI and multi-words for text classification. *Expert Systems with Applications, 38*(3), 2758–2765.

Part III
Monitoring

Chapter 5
Intelligent Web Data Management of Infrastructure and Software Monitoring

5.1 Introduction

5.1.1 Background

Along with the cloud's increasingly central role of the services in industry, deployed monitoring cloud applications and services as well as the applications is becoming a priority. Cloud monitoring may be viewed as a specialization of distributed computing monitoring and therefore inherits many techniques from traditional computer and network monitoring. However, as cloud computing environments are considerably more complex than those of legacy distributed computing, they demand the development of new monitoring methods and tools. Overall, the integration of cloud monitoring with related techniques to effect an end-to-end automated monitoring and provisioning process over cloud environments is a hitherto neglected research area.

In order to address these limitations, this chapter introduces intelligent Web data management of a lightweight hybrid cloud monitoring system. The monitoring system is composed of two layers in order to reduce the complexity: manager-agent monitoring of entity objects, and aspect-oriented cloud service monitoring. The first part of our monitoring framework is manager-agent monitoring. The manager is a separate entity that is responsible to communicate with the agent. Enabling the agent allows it collecting the monitoring data from the device locally. However, this approach applies even more to monitor the entity objects. For the monitoring of cloud Web services, this approach is inferior. Therefore, we propose another aspect-oriented cloud service monitoring approach as a supplementary item. Aspect-oriented monitoring approach is scattered by virtue of the functions of Web services, which can measure the quality of service (QoS) parameters of cloud Web service. Manager-agent and aspect-oriented monitoring approaches work in parallel, which constitute the whole monitoring framework.

© Springer International Publishing Switzerland 2016
K. Ma et al., *Intelligent Web Data Management: Software Architectures and Emerging Technologies*, Studies in Computational Intelligence 643,
DOI 10.1007/978-3-319-30192-1_5

5.1.2   Challenges and Contributions

Currently, third-party cloud Web services have spread among tenants and end users, providing a number of advantages over infrastructures. Although cloud providers claim a higher resilience, assessments of their actual availability are missing. Despite the advantages of cloud computing, small and medium enterprises in particular remain cautious while implementing cloud service solutions. The main drawbacks for the companies to adopt cloud computing as follows. First, companies lack qualified and trustworthy benchmarks to assess and monitor cloud services. Furthermore, companies lack approaches and metrics to adequately evaluate the quality of cloud services. The motivation of our framework is just the requirement of such monitoring system for a hybrid cloud.

Even though the cloud has greatly simplified the capacity provisioning process, it poses several new challenges in the area of Quality of Service (QoS) management. QoS denotes the levels of performance, reliability, and availability offered by a cloud application and by the platform or infrastructure that hosts it. QoS is fundamental for cloud users who uses cloud resources, such as infrastructure and software. Cloud users expect providers to deliver the advertised quality characteristics, and for cloud providers, who need to find the right tradeoffs between QoS levels and operational costs. However, finding optimal tradeoff is a difficult decision problem, often exacerbated by the presence of service level agreements (SLAs) specifying QoS targets and economical penalties associated to SLA violations.

We argue that there is no general cloud monitoring solution that may be applied to cover three major cloud service models: infrastructure as a service (IaaS), software as a service (SaaS), and platform as a service (PaaS) [1]. Cloud models are shown in Fig. 5.1. We choose to address the usage of cloud resources described by a set of quantitative metrics. These metrics are divided into three categories: IaaS metrics, SaaS metrics, and user experiences. As the PaaS models are emulated over an IaaS and SaaS base, we do not discuss it in this chapter. IaaS metrics include the performance and availability of physical hypervisor and virtual machines. While

Fig. 5.1 Cloud hierarchy models

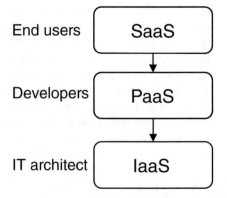

SaaS metrics include the performance of applications in the cloud, and the performance of Web services. We take user experience as an instructional factor, since it can help the tenants of the clouds to learn the user habits of end users. This is one of the effective ways to improve the quality of cloud service.

The contributions of this chapter are several folds. First, it presents the design of the hybrid cloud monitoring framework with open source solutions and extra significant development work. Second, the extension fine-grained mechanism of manager-agent monitoring is based on plug-ins bundles, which is an effective way to load custom plug-ins dynamically. Through a simple method built quickly from freely available parts, it is partially successful, suggesting this monitoring framework is used both in public and private clouds. Besides, aspect-oriented cloud service monitoring enables the developer to dynamically write references to aspects at join points and calculate the performance of cloud services. Aspects thus eliminate many lines of scattered code that the developers would otherwise have to spend considerable time in writing the monitoring codes.

5.2 Related Work and Emerging Techniques

5.2.1 Cloud Monitoring Categories

Figure 5.2 shows cloud deploy models: public, private, and hybrid cloud infrastructures [1]. Different cloud deployment models have different requirements. The differences are most distinct between private and public clouds. Private clouds, even if scalable, are limited to the resources owned by the operating organization. Regarding security, in private clouds, the data is under the organization's control, whereas public clouds require continual surveillance across multiple cyber-attack vectors. Public clouds often have geographically diffused large resource pools, which require more investment in monitoring traffic and ensuring scalability. Service metrics such as link availability and connection speed are essential information in public clouds but are of little value in private clouds. In public clouds, firewall settings may limit what can be monitored between cloud providers. Finally, public clouds need to provide monitoring information for clients, which requires the

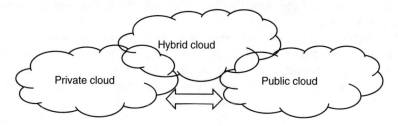

Fig. 5.2 Cloud deploy models

flexibility, customizability, and security. Hybrid cloud mixes the techniques from public and private clouds. The benefits and challenges are a combination of the items above. As a consequence, the topic of this chapter relates to hybrid cloud monitoring. In this chapter, we attempt to design and implement a mass of modules to support further monitoring of hybrid cloud.

5.2.2 Cloud Monitoring Methods

Monitoring of computing resources has become a hot topic of research interest and development for many years. However, monitoring in clouds faces many new challenges. First, due to the heterogeneity of components in the clouds, individual computer and network monitoring solutions and mechanisms need to be designed and implemented respectively, which is very costly. Second, much more monitoring concerns need to be covered in clouds than traditional monitoring software, such as multi-tenancy and hypervisors. Third, the large number of services, tenants, and end users in the clouds lead to the demand for a more intuitive and flexible way to implement monitoring in the cloud.

By exploring the recent literature, we discovered several monitoring systems, each one with its particular characteristics and abilities. Well-known clouds in the industry have their particular monitoring utilities. We will classify the available mechanisms and point out the drawbacks and inefficiencies [2].

Taxonomy 1: cloud monitoring level.
The first taxonomy is cloud monitoring level. A typical cloud architecture would include the levels: server, infrastructure, platform and application. On one hand, the lowest level is the server, which contains physical machines and network. On the other hand, the remaining three levels contain the virtual resources provided by tenants and end users. Current open-source monitoring tools are designed in the former way. There are plenty of tools for this taxonomy.

Nagios [3] is the industry standard in the infrastructure monitoring, offering complete monitoring and alerting for servers, switches, software, and services about the status of resources. It is designed to run checks on hosts and services using several external plug-ins and return the status information to administrative contacts. Although it includes valuable features and abilities, it does not provide a generic API and perform under small-time interval. The architecture of Nagios is shown in Fig. 5.3.

Eucalyptus [4] is open source computer software for building cloud computing environments. Systems that give users the ability to run and control entire virtual machines deployed across a variety physical resources. However, this tool is commonly referred to for IaaS only. The architecture of Nagios is shown in Fig. 5.4.

Ganglia [5] is a scalable distributed monitoring system for high-performance clusters and grid computing. It leverages widely used technologies and is at the base

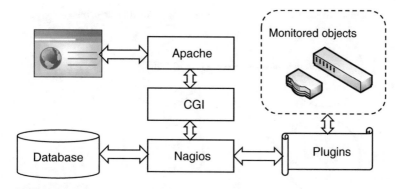

Fig. 5.3 Architecture of Nagios

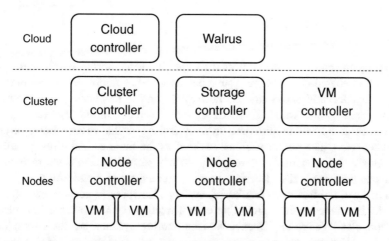

Fig. 5.4 Architecture of Eucalyptus

of a hierarchical design targeted at federations of clusters. The disadvantage of the method is not good for bulk data transfer (no windowed flow control, congestion avoidance, etc.). The architecture of Eucalyptus is shown in Fig. 5.5.

CloudSense provides a new switch design that performs continuous fine-grain monitoring via compressive sensing. This framework uses MapReduce straggler monitoring to report conventional status via the analysis and emulation.

Recently, some experts proposed a trust-based approach to make a node capable of finding the most reliable interlocutors. This approach avoids the exploration of the whole node space.

Taxonomy 2: cloud monitoring vision.
The second taxonomy is cloud monitoring vision. From a general perspective, client-side and cloud-service-provider-side monitoring can be distinguished. These

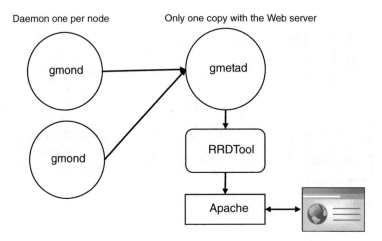

Fig. 5.5 Architecture of Ganglia

two complementary visions address different cloud monitoring requirements, creating differentiated views of the system behavior and evolution. There are plenty of approaches for this taxonomy. Some experts find a monitoring framework for clouds service side, which succeeds in the scalability of the monitoring. This framework is spread in different layers (service, virtual environment, physical resources, etc.). The proposed framework offers the libraries and tools to build its own monitoring system. However, they do not present performance metrics and point out the boundaries. Besides, there are commercial cloud solutions, which often make use of their own monitoring systems. However, in general these systems have limited functionality, providing only a fraction of the available information. Amazon cloudWatch and OpenNebula monitoring systems are the examples of this kind. For the private cloud, it is easy to monitor the client-side and cloud-service-provider-side, since we have total control of it. For the public cloud, we provide only client-oriented monitoring. Those cloud monitoring levels and visions motivated us to implement a lightweight framework that monitors hybrid clouds. To this end, the design of a monitoring system should keep up with the expansion of the flexible ability, when an application or infrastructure scales up or down dynamically.

Dixon [6] envisioned monitoring systems with interchangeable components focused on a single responsibility. Such a system architecture should show the following characteristics: resilient, self-service, automated, correlative and craftsmanship. As a result, the support of the monitoring entity objects in our framework depends on the plug-ins.

5.2.3 Cloud Monitoring Methods

There are several famous monitoring tools. Round Robin Database Tool (RRDtool) [7] is the open-source industry standard, high-performance data logging and graphing system for time-series data. The data analysis part of RRDtool owns the ability to create graphic representations of the data values collected over a definable time period. However, the current plug-ins of RRDtool can not support the cloud monitoring, such as the hypervisors. Collectd gathers statistics about the system it is running on and stores this information. Those statistics can then be used to find current performance bottlenecks and predict future system load. With these monitoring tools, it is possible to deploy a hybrid cloud monitoring solution using extra significant development work. Although these tools are not designed for cloud monitoring, the refactor of them may adapt to the cloud. The architecture of RRDtool is shown in Fig. 5.6.

5.2.4 Cloud Web Service Monitoring in the Cloud

For the service in the cloud, the above-mentioned monitoring methods are not applicable because of the heterogeneous content. A runtime model for cloud monitoring (RMCM) is proposed to denote an intuitive representation of a running cloud by focusing on common monitoring concerns. Raw monitoring data gathered by multiple monitoring techniques are organized by RMCM to present a more

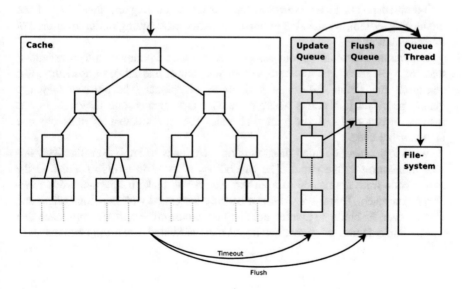

Fig. 5.6 Architecture of RRDtool

intuitive profile of a running cloud. In the SaaS layer, it monitors the applications with respect to their design models and required constraints. For this issue, it converts the constraints to a corresponding instrumented code and deploys the resulting code at the appropriate location of the monitored applications. This makes RMCM an invasive approach, since it modifies the source code of the applications. Some experts propose a monitoring architecture for cloud computing. It describes a Quality of Service (QoS) model that collects QoS parameter values such as response time, cost, availability, reliability and reputation. Although their architecture is interesting, the implementation of this architecture is not clear.

5.2.5 Aspect-Oriented Programming

Aspect-oriented programming (AOP) [8] is based on the idea that computer systems are better programmed by separately specifying the various concerns (properties or areas of interest) of a system and some description of their relationships, and then relying on the mechanisms in the underlying AOP environment to weave or compose them together into a coherent program. The key difference between AOP and other approaches is that AOP provides component and aspect languages with different abstraction and composition mechanisms. A special language processor called an aspect weaver is used to coordinate the co-composition of the aspects and components. Modularized crosscutting concerns are called aspects. An aspect, if it can not be cleanly encapsulated in a generalized procedure. AOP distinguishes between two approaches. Static crosscutting affects the static type signature of a program, whereas dynamic crosscutting allows intercepting a program at well-defined points in its execution trace. After evaluating the flexibility of the monitoring system, we decided to pursue dynamic crosscutting option over ease of implementation.

In dynamic crosscutting, join points are well-defined points within an execution trace of a program. Pointcuts are sets of join points and advice are method-like constructs that define the behavior of these join points. The pointcut language defines abundant join points where the advice is integrated into the code. In this manner, aspect-oriented programming eases the development of reusable and maintainable code.

We adopt aspect-oriented programming techniques to implement the Web service monitoring in the cloud. The capabilities of AOP in terms of isolating the aspect code from the source code of the used server make it a non-invasive monitoring approach. This framework can be fully integrated with the cloud client and cloud server. It should be pointed out that this framework is a generic approach that can be implemented in several heterogeneous distributed monitoring contexts.

5.3   Requirements

5.3.1   *Hierarchy of Resource Entity Models*

Concrete resource entities in a running hybrid cloud are organized in a hierarchy, as depicted in Fig. 5.7. This hierarchy can be extended by adding new entities emerging in different cloud layers [9]. Such an organization makes an extension to the model more feasible. New entities can be added in directly by inheriting one of the existing entities. Each entity consists of a set of key/value pairs. Children entities can inherit attributes from their parents. And values are obtained from real-time cloud monitoring statistics. Based on the entities, we design the monitoring system. For further monitoring, we can extend the hierarchy of resource models.

5.3.2   *Requirements of Monitoring*

We have the following requirements while designing a cloud monitoring architecture.

Monitoring of SaaS objects: We want to monitor different kinds of SaaS objects, such as software, Web services.
Monitoring of IaaS objects: We want to monitor different kinds of IaaS objects, such as virtual machines, hypervisors, and physical machines.
Monitoring of user experience: We want to monitor user experience to guide the cloud service providers to supply good service.

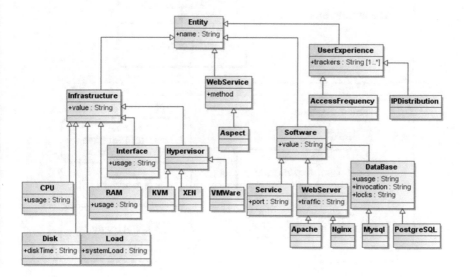

Fig. 5.7 Hierarchy of entity models

5.4   Architecture

5.4.1   Cloud Monitoring Architecture

This section presents the architecture of cloud monitoring adapted to any cloud infrastructure, which is presented in Fig. 5.8. The architecture is divided into four main components: the first one, regarding the IaaS monitored object, the second one, regarding the SaaS monitored object, the third one, regarding the monitoring access, and the last one, regarding the data gathering. Since the monitoring of Platform as a service (PaaS) depends on the specific platform, we do not discuss it in this chapter.

At the bottom layer lies the physical resources, such as the hardware resource utilization (CPU, memory, disk, networking et al.).

The second layer from the bottom is the IaaS monitored object, including the physical machines, the hypervisors that manage the guest operating systems, and the virtual machines. They are all the monitored objects in IaaS level. This performance metrics, which stays the same regardless of distinguished infrastructures, are extracted to model the runtime infrastructure in the cloud.

The third layer from the bottom is the SaaS monitored object, including the applications, the Web servers and the Web services. They are all the monitored objects in SaaS level. These applications provide a wide set of services that assist

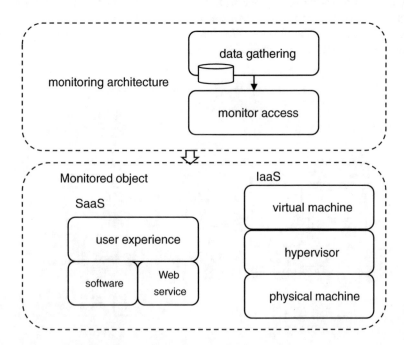

Fig. 5.8 Requirements of cloud monitoring architecture

developers in delivering a professional and commercial service to end users. Commonly, the software in the SaaS layer will expose their own interface or API for monitoring and management, and this metric is gathered and organized to form the runtime monitoring model of this layer. Besides, the user experience is also an important entity to monitor. This point is generally omitted by the traditional monitoring tools. We put the user experience of cloud tenants as a reference to discover the relationship between users' habits and cloud performances. Also it is the guidance on the future management of the cloud. Interactive information and action between end users and clouds are monitored. It mainly contains the frequency of access, IP distribution, time of staying and so on. The service developers may adapt their software to attract more users of hybrid clouds.

The second layer from the top is monitoring access, which provides an independent way to access the monitored information from the bottom. It is an abstraction layer that provides a unique view of both physical and virtual cloud monitored objects, regardless of its different objective, composition and structure. This layer accesses both the virtual and physical systems.

At the top is data gathering, which provides the storage for the monitoring information obtained in the previous layer. It includes a historical archive of the evolution of the different cloud objects. Since each user has its own cloud vision and monitoring level, this layer does not only store monitoring data but it coordinates the user queries. In this way, each user can access the monitoring level and vision required.

5.4.2 Manager-Agent Architecture

In a cloud, multiple entities need to be monitored simultaneously to coordinate their resource utilization to achieve a balance. Therefore, the framework is required to monitor the resources individually and take decisions centralized. This chapter achieves a manager-agent styled monitoring framework, as shown in Fig. 5.9. The manager provides the interface between the human network manager and the management system. While the monitoring agent is a small daemon which collects system information periodically. In order to analyze the collected data, we provide the mechanisms to store and monitor the values in a variety of ways. These agents are responsible for collecting runtime information and sending UDP packet to the manager. By default, we provide the basic monitoring of common metrics: IaaS monitoring plug-ins, and SaaS monitoring plug-ins. Other plug-ins are written to allow the developer to further extend the default monitoring system. All the personalized plug-ins conforming to our interface specification can be remotely deployed and installed without requiring a reboot in the forms of bundles. We provide the user interface with the administrators and tenants of hybrid clouds. This will help them view the operating condition of hybrid clouds.

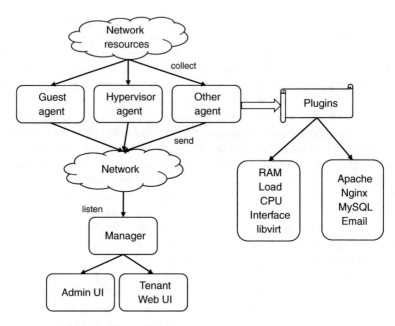

Fig. 5.9 Manager-agent architecture and its plug-ins

5.5   Evaluation

For the IaaS monitored objects, we implemented the monitoring of the physical machine, the hypervisor, and the virtual machine.

- The monitoring of the physical machine presents an overall of metrics for the whole cloud.
- The monitoring of the hypervisor presents an overall of metrics for the guest virtual machines without the access to the internal machines.
- The monitoring of the guest virtual machine presents an overall of metrics for the internal machines separately.

For the SaaS monitored objects, we implement the monitoring of the software, the Web service, and the user experience.

The proposed monitoring framework provides the comprehensive monitoring in the forms of different plug-ins. These plug-ins include virtualization, availability, performance via SNMP, user experience tracker, and over-commit monitoring. It provides many kinds of alert and information statistics to help the administrator find the problems in real time. Each module is a daemon, which collects the data from the monitored object.

5.5.1   Virtualization Monitoring

This monitoring module uses the libvirt API to gather the statistics of the guests on a system shown in Fig. 5.10. The libvirt public APIs support many commonly hypervisor drivers, such as KVM, XEN, Hyper-V, VMware ESX and VirtualBox. With libvirt plug-in, CPU, memory, networking and device usage for each guest could be collected without installing any software on the guest. As the module receives statistics from the hypervisor directly, it is suitable for the physical and virtualized machines. The system load collected from the last one week is shown in Fig. 5.11. Due to the delegation to one or more internal driver, the libvirt public API reveals the ability to support the new hypervisor.

5.5.2   Service Availability Monitoring

Availability monitoring module can refer to the process of collecting data and reporting on the status of the critical website. In other words, it means the availability for the WWW service in the cloud. In order to get the accurate availability, the probe points are distributed around the world. First, we deploy some probe points in the backbone network of the popular Chinese Internet Service Providers (ISPs): China Unicom (CNC), China Telecom (CTC), and China CERNET. Second, we deploy some probe points in the cloud host (in USA, UK, Australia, etc.) that is purchased from the Amazon Elastic Compute cloud (EC2) service. We use tcping command to determine the connectivity over TCP. Applying this monitoring module, the detail of the availability curve graph of a virtual machine in hybrid cloud is shown in Fig. 5.12. The notification system notified administrators of target availability status changes according the incident rulesets.

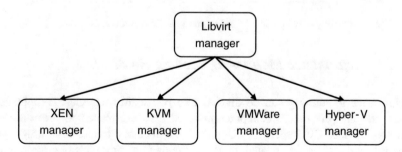

Fig. 5.10 Infrastructure libvirt monitoring architecture

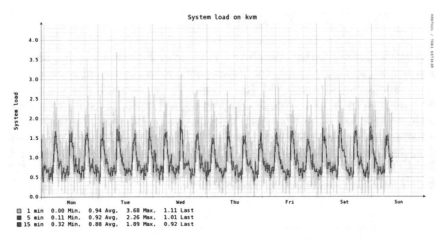

Fig. 5.11 The system load collected from the last one week

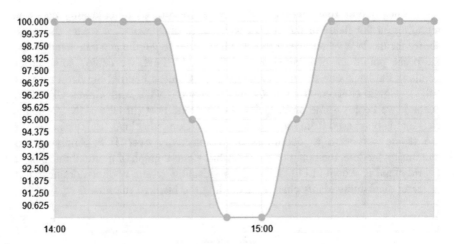

Fig. 5.12 Availability curve graph of a virtual machine in hybrid clouds

5.5.3   *Performance Monitoring Module via SNMP*

Performance monitoring can be divided into two main categories: computation-based and network-based. Computation-based tests are related to the following metrics: service, CPU speed, CPU utilization, memory usage, disk usage, etc. Network-based tests are related to the following metrics: network throughput, packet loss, bandwidth, etc. We implement the manager-agent monitoring module to measure these metrics using Simple Network Management Protocol (SNMP). Furthermore, we present the statistics graphing of historical reference of these metrics in the browser. This module will appear an alert when the thresholds of the

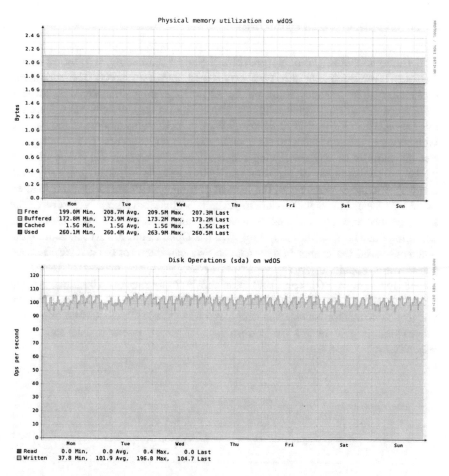

Fig. 5.13   An example of the memory and disk monitoring in hybrid clouds

monitored objects were exceeded. Services and processes running on devices can
be monitored to verify whether they are running or not. An example of the memory
and disk monitoring in hybrid cloud is shown in Fig. 5.13.

In our solution, we use Java SNMP package API to implement the SNMP
monitoring manager, which is responsible to communicate with the SNMP agent.
We take advantage of SNMP v2c API to get responses from agents. In our solution,
we deploy the open-source SNMP agent in the physical machine, the hypervisor
and the virtual machine to collect the monitoring data. The extensible agent for
responding to SNMP queries to management information includes built-in support
for a wide range of the Management Information Base (MIB) information modules,
and can be extended using dynamically loaded modules, external scripts and
commands.

5.5.4 Application Monitoring

Application monitoring module provides the most commonly monitoring in the software layer, including the Web server and database server. The monitored application often provides some public APIs. Regarding different Web servers, there are different monitoring methods, depending on the different APIs. Here are two examples of Web server monitoring. We use mod_status module of the Apache Web server to to keep watch on the operation of the server serves. The details given are: the number of worker serving requests and idle worker, the status of each worker, a total number of accesses and byte count served, averages giving the number of requests per second, the current percentage CPU used by each worker, etc. We provide stub_status module to get some status from Nginx Web server. The active connections and handled requests are calculated by this API. Here is an example of database server monitoring. We use Open Replicator API to capture the statement-based log events as they happen. We can intercept the database request (alter, select, update, insert and delete) to be analyzed. The number of issues per second for various SQL-commands are shown in Fig. 5.14. The Java method capture is shown as follows, which is the implementation of interface Capture. We use Java method selectMonitoring, insertMonitoring, updateMonitoring and deleteMonitoring to record the performance of select, insert, update and delete respectively.

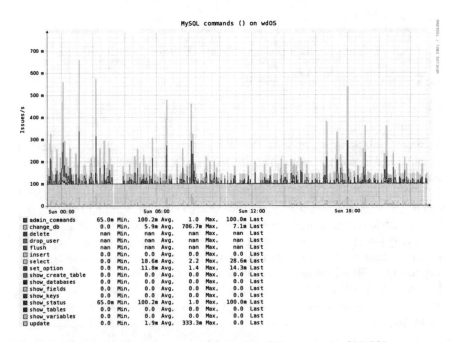

Fig. 5.14 The number of issues per second for various SQL-commands of MySQL

5.5.5 User Experience Tracker

Compared with the availability monitoring, user experience tracker provides a more accurate method to calculate the access speed and position real experience of each visitor. This module embeds a JavaScript in the webpage of the monitored website. The website in clouds acts as the tenant, and the visitor of the website acts as the end user. The user access information (IP and location) is automatically recorded after the load of the webpage. For example, a multi-tenant social website is served for many cities. Although the visitors in different cities access the different entrances of the website, they actually visit the same website. Figure 5.15 shows the average response time of visitors of the website, which reflects the quality of our website. As depicted in Fig. 5.15, the average response time of NingXia is the longest, while the response time of Tibet is the shortest. We can conclude that the visitors in NingXia have poor user experience, and the visitors in Tibet have a better user experience. That is the guidance for the cloud owners to optimize the network. It plays an important part in discovering the relationship between end user habits and cloud performances. Furthermore, the metrics from the tracker is a indirect guidance on how to attract more users in the hybrid clouds.

5.5.6 Over-Commit Monitoring

In the area of IaaS, over-commit refers to the practice of committing more virtual resources to customers than the actual resources available on the underlying physical cluster. Most of the prevailing hypervisors, such as Hyper-V, VMWare ESX, KVM, and XEN, support both CPU and memory over-commit. When an IaaS service provider practices over-commit, the over-commit parameters are usually unknown to the end user. When an end user creates a VM that is labeled as 1 CPU core and 1 GB memory, the CPU usage limit for that particular VM might be 1, 0.5 or even 0.1 physical processor core. Similarly, a host server with 16 GB physical memory may be committing 24 GB or even 32 GB virtual memory to virtual machines. Therefore, we provide over-commit monitoring with the cloud tenants, which help the tenants find out the actual performance of the virtual machines. We have built in a local

Province	Tracker times	Response time	
Ningxia	125	150.6	
Gansu	146	98.2	
Shanxi	123	62.4	
Sichuan	116	52.3	
Sinkiang	128	45.1	
Tibet	145	28.6	

Fig. 5.15 An example of user experience tracker

interface in the guest virtual machines that is served for cloud tenants. This interface is implemented in open-source benchmark suite UnixBench, which provide a basic indicator of the performance of a Unix-like system.

The experiments show that the system scores of UnixBench rather than the parameters of CPU core, and the amount of memory indicate the actual performance of the virtual machine in the same hypervisor. Figure 5.16 shows the UnixBench test results of virtualized guest. For virtual machines in the same hypervisor, as the virtual machine gets bigger, the performance gets better in the none over-commit circumstance; for virtual machines from the same hypervisor, as the virtual machine gets bigger, the performance gets worse in the over-commit circumstance.

5.6   Discussions

As depicted from Fig. 5.17, there are two objects. The first object is the cloud service client, which is located at client side. From this point of view, monitoring information of this type helps the tenant client to understand the QoS received and optimize their use. The second object is the cloud service server, which is located in cloud-service-provider side. From this point of view, monitoring information of this type provides the knowledge about the internal functioning of the different cloud

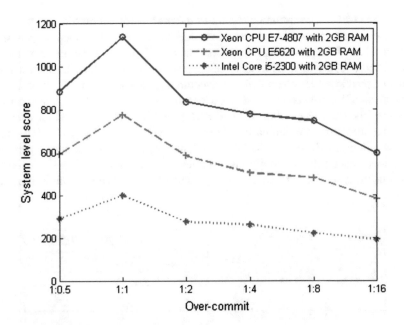

Fig. 5.16 UnixBench test results of virtualized guest

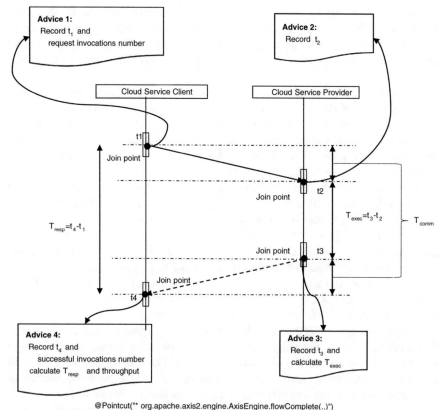

@Pointcut("* org.apache.axis2.engine.AxisEngine.flowComplete(..)")

Fig. 5.17 QoS parameters and the sequence diagram

objects with the cloud administrators in order to guarantee QoS. It can be also used as usage and performance log to optimize the service provided.

This framework measures five QoS parameters: the execution time, the response time, the communication time, the throughput and the availability. They are explained below. All the terminology is shown in Fig. 5.17.

The communication time T_{comm} is the time needed to transfer the request from the client to the server plus the time needed to transfer the response from the server to the client. It is the sum of the two parts. Thus, the response time shown as follows is the sum of the time necessary for executing the request and the communication time.

$$T_{resp} = T_{comm} + T_{exec}$$

The execution time T_{exec} measures the time needed to execute a request on the server.

The response time T_{resp} defines the time needed to serve a request. It starts when the client sends its request and finishes when the client receives the corresponding response. As a result, it is the temporal difference between the instant (t1) when the client invokes the request and the instant (t4) when the client receives the corresponding response.

The throughput measures the number of successful requests (SuccRequests) during a period of time T. In other word, the throughput is measuring the success number of requests per second. Throughput = SuccRequests/T, where SuccRequests represents the number of successful requests during a period T. T is a parameter that is set when this framework is configured. We mean that the successful request is with a successful response reached the server.

The availability measures the accessibility of a service. It is calculated using the formula shownas follows. Availability = SuccRequests/AllRequests, where AllRequests is the number of all requests sent during the period T.

Our framework is based on the aspect-oriented programming code that intercepts the methods of the client and server at well defined join points to collect data and measure these parameters at important instants of time. These instants are t1–t4, where t1 is the instant when the client invokes the request, t2 is the instant when the server receives the request, t3 is the instant when the server sends the response, and t4 is the instant when the client receives the response.

The proposed aspect code computes the number of request invocations by advice 1. At the mean time, the unique identity is also recorded. Generally, the unique identity is composed of the IP address of the client and the client session. In addition, the execution time is calculated by advice 3. Moreover, the number of successful sent requests and response time is evaluated by advice 4.

Recorded timestamps help us to calculate QoS parameters. The procedure is as follows: First, our framework creates four specific join points. Each one corresponds to an instant (t1, t2, t3 and t4) intercepting method calls. Second, it executes the corresponding advice (for each join point) which consists of recording the timestamp and calculating the number of invocations at the client (advices 1 and 4). When all instants have been processed, it computes the difference between t4 and t1 to deduce the response time. It also subtracts t2 from t3 to calculate the execution time. The difference between the response time and the execution time represents the communication time. It assesses the throughput by calculating the number of successful invocations at the client side and dividing this number by a period of time T.

The aspect-oriented cloud service monitoring has been implemented within Axis2. The implementations include the client and server of AOP service monitoring.

5.6.1 *Monitoring Client of AOP Service*

```
@Aspect
public class MonitorClient {
  @After("execution(* org.apache.axis2.engine.AxisEngine.invoke(..)")
  public void advice1(){
    t1 = System.nanoTime();
    AllRequests++;
  }
  @After("execution(* org.apache.axis2.engine.AxisEngine.flowComplete(..)")
  public void advice4(){
    t4 = System.nanoTime();
    tresponse = t4 - t1;
    SuccRequests++;
  }
}
```

The first component of our AOP monitoring approach is the monitoring client, which is the aspect-oriented advice 1 and 4 that intercepts the client at t1 and t4. Its implementation is based on the identification of the methods that the service engine invokes at t1 and t4. In Axis2, the method invoked at t1 is invoke of the AxisEngine class located in org.apache.axis2.engine package. When the client sends a request, the Axis engine on the client side is invoked via the method invoke(…). At the instant t4, the method flowComplete of the AxisEngine class located in org.apache.axis2.engine package.

5.6.2 *Monitoring Server of AOP Service*

```
@Aspect
public class MonitorServer {
  @After("execution(* org.apache.axis2.engine.AxisEngine.invoke(..)")
  public void advice2(){
    t2 =System.nanoTime();
  }
  @After("execution(* org.apache.axis2.engine.AxisEngine.flowComplete(..)")
  public void advice3(){
    t3 = System.nanoTime();
    texec = t3 - t2;
  }
}
```

The second component of our AOP monitoring approach is the monitoring server. It is the aspect-oriented advice 2 and 3 that calculates the execution time. To implement the monitoring server using Axis2, it is necessary to identify the pointcut 2 and pointcut 3. Pointcut 2 corresponds to the instant t2 when the request arrives at the server side, and pointcut 3 describes the instant t3 when the response leaves the server side. The execution of the method invoke(…) of the class AxisEngine located in the org.apache.axis2.engine package is in charge of the request processing at the server side. This means that recording the instants before and after the execution of this method lead to the computation of the execution time value. Thus, pointcut 2 and 3 correspond to the interception of the execution of this method, and the related advices should be applied before and after this method, respectively.

AOP service monitoring also handles multiple clients, which is able to distinguish between clients by using their IP addresses. Furthermore, AOP service monitoring uses the request session ID to differentiate between concurrent requests running on the same client. Therefore, it associates monitored QoS parameters to the corresponding client. Based on AOP, the monitoring server extracts the client and the server IP addresses corresponding to the collected monitored QoS parameters. Furthermore, the monitoring client distinguishes between the concurrent requests running on the same client, while extracting their session IDs. It should be pointed out that the AOP monitoring components are completely independent of the original server. In fact, the developed aspect codes are not located inside the server source code. With the help of the weaving mechanism of AOP, they intercept the methods at defined join points and record the relevant timestamp information without modifying the source code of the monitored service.

5.7 Conclusions

Nowadays, monitoring plays an important part in hybrid clouds. However, current monitoring solutions on the popular metrics of the hypervisors, the virtual machines, and the physical machines in hybrid clouds are minimal at best. Towards our goal of building such a lightweight and scalable architecture, we integrate some open-source monitoring tools with some extra significant secondary development work to implement some modules. First of all, we introduce the monitoring architecture and hierarchy of resource entity models. User experience is pointed out as a reference to discover the relationship between users' habits and cloud performance. In addition, we discuss the manager-agent and aspect-oriented architecture to perform end-to-end measurements at both virtual and physical machines and software in hybrid clouds. By default, we provide the basic monitoring of common metrics. Each monitoring is implemented in the forms of plug-ins. Aspect-oriented cloud service monitoring allows the developer to dynamically write references to aspects at join points to calculate the performance of cloud services. Thus, they eliminate many lines of scattered code. Finally, the implementations of this framework and some experiments have demonstrated the high performance of the proposed framework.

References

1. Armbrust, M., Fox, A., Griffith, R., Joseph, A. D., Katz, R., Konwinski, A., & Zaharia, M. (2010). A view of cloud computing. *Communications of the ACM, 53*(4), 50–58.
2. Aceto, G., Botta, A., De Donato, W., & Pescapè, A. (2013). Cloud monitoring: A survey. *Computer Networks, 57*(9), 2093–2115.
3. Barth, Wolfgang. (2008). Nagios. No Starch Press.
4. Nurmi, D., Wolski, R., Grzegorczyk, C., Obertelli, G., Soman, S., Youseff, L., & Zagorodnov, D., et al. (2009, May). The eucalyptus open-source cloud-computing system. In *9th IEEE/ACM International Symposium on Cluster Computing and the Grid, 2009. CCGRID'09.* (pp. 124–131). IEEE.
5. Ganglia. http://ganglia.sourceforge.net.
6. Dixon. https://www.dixonvalve.com/.
7. Oetiker, Tobias. (2005). "RRDtool".
8. Elrad, T., Filman, R. E., & Bader, A. (2001). Aspect-oriented programming: Introduction. *Communications of the ACM, 44*(10), 29–32.
9. Ma, K., Sun, R., & Abraham, A. (2013). Toward a module-centralized and aspect-oriented monitoring framework in clouds. *Journal of Universal Computer Science, 19*(15), 2241–2265.

Chapter 6
Intelligent Web Data Management of WebSocket-Based Real-Time Monitoring

6.1 Introduction

6.1.1 Background

Wireless sensor networks (WSNs) [1] are used in many industrial applications, such as industrial process, environmental and health monitoring, and so on. With the development of green computing and Internet of Things, more and more people have contributed an increased demand to reduce the energy consumption of buildings [2]. This implies the necessity of a conscious way of thinking and actions regarding efforts to monitor smart objects in remote intelligent buildings.

WSNs used in intelligent building management systems consist of different types of sensor nodes measuring metrics such as temperature, humidity, light and smoke. In addition, the systems may include actuators, gateways, servers and communication and application software on different levels as well as different home appliances. A large amount of research has been conducted focusing on different aspects of WSN for monitoring. There are some deficiencies of current approaches, such as the delay and concurrence of the monitoring. Some emerging techniques, such as HTML5 WebSocket, can assist this monitoring.

6.1.2 Challenges and Contributions

However, there are some challenges of current monitoring for remote intelligent buildings in two aspects.

The first aspect is the monitoring approach. The intelligent buildings need a real-time monitoring approach with fast loading and low latency. Moreover, high concurrency and low consumption play an important role in the monitoring because sensor node or server has limited energy and computational resources.

© Springer International Publishing Switzerland 2016
K. Ma et al., *Intelligent Web Data Management: Software Architectures and Emerging Technologies*, Studies in Computational Intelligence 643, DOI 10.1007/978-3-319-30192-1_6

The second aspect is the storage of the historical monitoring data. In the era of big data, the historical monitoring sensor data is so huge that the relational database will encounter the bottleneck problems in the track of historical data. In addition, there are many redundant data in the historical data warehouse.

Future monitoring system will collect a large amount of real-time monitoring data. In the context of monitoring of remote intelligent buildings, eventual consistency is acceptable. The features of the track of historical monitoring data determine that monitoring systems exhibit a high read to write ratio. Therefore, we can utilize emerging NoSQL [3] instead of Relational Database Management Systems (RDBMS).

Our motivation of this chapter is trying to address these two limitations to propose a WebSocket-based real-time monitoring architecture for remote intelligent buildings. We extend the Internet of Things paradigms to support more scalable and interoperable and monitoring the sensors for intelligent buildings in real-time and bi-directional manner.

The contributions of this study are divided into two aspects. The first aspect is our WebSocket-based monitoring approach. First, this WebSocket-based monitoring approach is pervasive in other high-frequency application scenario. Second, the WebSocket-based monitoring approach is advantageous because it provides browser independence with low latency and low-energy. Third, monitoring data in the transmission is updated in the browser in the high frequency, without the use of any browser plug-ins. Our solution also provides WebSocket emulation to support those popular legacy browsers.

The second aspect is the storage model of historical monitoring data. First, we propose the storage model with lifecycle to reduce redundant data and storage space. Second, we utilize the NoSQL data warehouse to provide the high-performance tracking of the historical monitoring sensor data.

6.2 Related Work and Emerging Techniques

Before we introduce our WebSocket-based real-time monitoring architecture for remote intelligent buildings, we discuss the current networking technology of intelligent building, the classical monitoring methods, and the storage model of monitoring data.

6.2.1 Networking of Intelligent Building

Many efforts are currently going towards networking smart things of intelligent buildings (e.g. Radio Frequency Identification (RFID [4]), wireless sensor networks, and embedded devices) on a larger scale. Compared with WSNs themselves, Internet of Things has mainly focused on establishing connectivity in a variety of challenging and constrained networking environments and the next logical objective is to build on top of network connectivity by focusing on the application layer.

There are some key issues of the networking of intelligent buildings. First, how to utilize the emerging Web technology to assist the monitoring of smart objects in remote intelligent buildings is challenging. The new sensor Web is also associated with a sensing system which heavily utilizes the World Wide Web (WWW) to define a suite of interactive interfaces and communication protocols abstracting from well-accepted and understood standards (such as Ajax, REST, HTML5, etc.), especially well-suited for environmental monitoring.

Second, how to support different heterogeneous network (such as ZigBee and 6LoWPAN) is challenging. Today, WSN system technologies suffer from non-flexibility and proprietary solutions. It is necessary to come to an understanding and bridging the gap of different protocol stacks. There are two popular wireless communication protocols adopted by IEEE 1451.5 standard: ZigBee and 6LoWPAN. However, there are some deficiencies of these two protocol stacks. Although ZigBee has the advantage of power saving and low cost, it is rather immature compared with Internet Protocol (IP) which has been developed over the past 40 years. The main disadvantages of ZigBee include short range, low complexity, and low data speed. 6LoWPAN, an alternative to ZigBee, is an acronym of IPv6 over Low power Wireless Personal Area Networks. As shown in Fig. 6.1, at the physical layer and the data link layer, it uses the same IEEE 802.15.4 protocol as ZigBee. 6LoWPAN has defined encapsulation and header compression mechanisms that allow IPv6 packets to be transferred over WSN. Even if there are a large number of devices deployed in a WSN, each device can also be assigned with a unique IP address. This feature makes it easy to support end-to-end communication. However, this protocol stack does not define a specification for the layers above IP. Both ZigBee and 6LoWPAN networks cannot communicate with each other without a sensor gateway. The networking approach is the integration of wireless Ethernet and WSNs (such as ZigBee and 6LoWPAN).

Fig. 6.1 ZigBee/6LoWPAN protocol stack

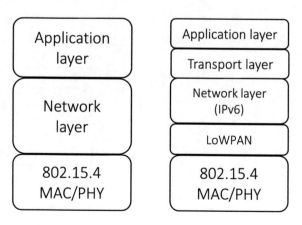

6.2.2   Classical Monitoring Methods

The most common monitoring is Web monitoring, which is suited for remote intelligent buildings. Currently, there are several categories of methods of monitoring [5, 6].

The first monitoring approach is HTTP long polling (hereinafter referred to as HTTP polling), which is shown in Fig. 6.2. HTTP polling is a variation of the traditional polling technique, but it allows emulating a push mechanism under circumstances where a real push is not possible, such as sites with security policies that require rejection of incoming HTTP Requests. However, polling has many disadvantages, including unnecessary requests and the response can be delayed as much as the polling interval.

The second monitoring approach is primitive Socket in ActiveX, which is shown in Fig. 6.3. The Socket control components provide easy-to-use Socket ActiveX control to develop applications that communicate using either TCP or UDP protocols. These have the advantage of working identically across the browsers with the appropriate plugin installed and need not rely on HTTP connections, but the disadvantage of requiring the plugin to be installed. Besides, ActiveX runs from Windows Internet Explorer only, and embedding application in the browser would affect browser-side performance.

The third monitoring approach is FlashSocket, which is shown in Fig. 6.4. FlashSocket relays make use of the XMLSocket object in a single-pixel Adobe Flash movie. The advantage of this approach is that it appreciates the natural read-write asymmetry that is typical of many web applications and as a consequence it offers high efficiency. Since it does not accept data on outgoing sockets, the relay server does not need to poll outgoing TCP connections at all, making it possible to hold open tens of thousands of concurrent connections. In this model, the limit to scale is Flash itself. Web applications that feature Flash features tend to take longer to load than those that do not.

Fig. 6.2 HTTP long polling

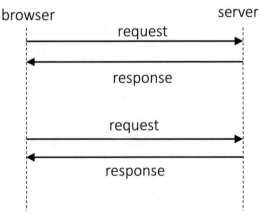

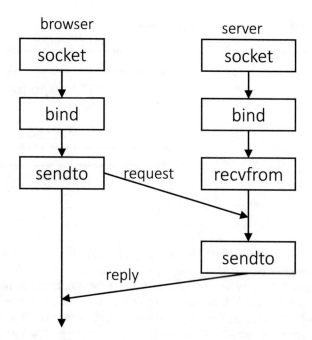

Fig. 6.3 Socket in ActiveX

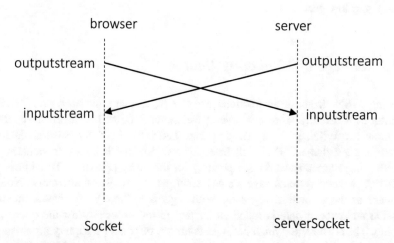

Fig. 6.4 FlashSocket

The last monitoring approach is WebSocket, which is shown in Fig. 6.5. WebSocket is a web technology in the HTML5 specification providing full-duplex communications channels over a single TCP connection. In this way, the WebSocket protocol makes possible more interaction between a browser and a web server. However, the legacy browsers lack of the support of the emerging WebSocket technology.

Fig. 6.5 WebSocket

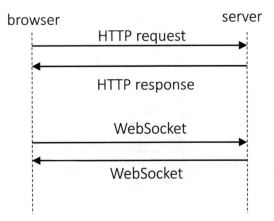

There are many drawbacks of the above approaches. We will deeply analyze the monitoring approaches. Our investigation explored the WebSocket compatible server integrated with WSNs to support every browser. Even if WebSocket is selected as the transport, WebSocket compatible server does more than WebSocket itself, including heartbeats, timeouts and disconnection features. This networking has led to a greater awareness of the conditions of buildings. The main benefits of this are: economies of scale gained from real-time monitoring, fast loading, low latency and low cost.

6.2.3 Storage of Monitoring Data

Existing WSN database abstractions use the SQL query interface and RDBMS techniques. Currently there are several traditional RDBMS database abstractions available for WSN, such as TinyDB and TikiriDB. Recently, several database experts argued that RDBMS will have the I/O bottleneck issues encountered by handling huge amounts of data, especially for the query operation. To address this limitation, several systems have already emerged to propose an alternative NoSQL database as the storage of big data. With regards to NoSQL databases, handling massive amounts of data is much easier and faster, especially for the query processing. The advantages which NoSQL databases offer over the query processing is reflected when it exhibits a high read to write ratio. Currently there are several NoSQL databases available, such as HBase and BigTable.

Although there are some best solutions of NoSQL, few publications have mentioned the data warehouse technologies about NoSQL. For the monitoring system, the basic business is the track of huge historical data. Therefore, we utilize MongoDB as the storage of historical monitoring data in WSNs. The motivations for this approach include high availability and query efficiency.

Fig. 6.6 Record with timestamp

Fig. 6.7 Record with versions

pk	a1	...	an	versions
				1
				2

Current storage models of RDBMS data warehouse provide lessons on the NoSQL stores. The first historical storage model is called record with timestamp [7], which is shown in Fig. 6.6. In this solution, the historical data are saved in another separate table with the timestamp every day. However, this solution has too many disadvantages. The first one is that the data warehouse is huge enough so that the storage becomes a disaster over a long period. The second one is that it will consume too many excessive storage spaces due to the redundant data.

The second historical storage model called record with versions [7], which is shown in Fig. 6.7. In this solution, all the historical data are tracked by creating multiple records for a given natural key in the dimensional tables with separate surrogate keys and/or different version numbers. This approach is suitable in the frequent updates scenario. However, the historical monitoring data are new data that never be altered.

6.3   Requirements

We have the following requirements while designing a WebSocket-based real-time monitoring architecture.

Pervasive real-time monitoring: We want to provide a general real-time monitoring architecture with some features: high-frequency, low latency, and low-energy.

Browser independence: We want to provide a Web-based monitoring, which is independent of the specific browsers.

Historical storage model: We want to provide a historical storage model to store monitoring data with low redundancy and space.

6.4   Architecture

6.4.1   Overview of System Architecture

The core idea of the proposed system is to enable wireless Web-based interactions with sensor gateways and WebSocket and Web server for monitoring the remote intelligent buildings. Figure 6.8 shows the proposed WebSocket-based monitoring system for WSNs of remote intelligent buildings in WSNs. This WebSocket-based monitoring system is composed of three points of view in order to reduce the complexity: WSNs of remote intelligent buildings, WebSocket-based real-time monitoring, and storage of the monitoring sensor data.

6.4.2   WSN of Intelligent Buildings

We seek to integrate WSNs with Ethernet communications to support data monitoring. This system is piloted by the deployment of WSNs in a remote intelligent building using the communication standards ZigBee and 6LoWPAN. These standards are embedded into a large number of chips for building automation. In our

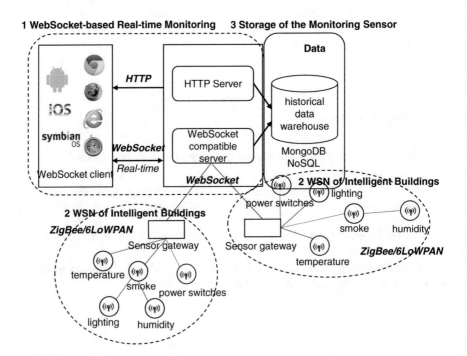

Fig. 6.8 Monitoring system architecture

solution, we adopt these two protocol stacks to support more sensor devices. We provide two types of sensor networks (ZigBee and 6LoWPAN) to collect the monitoring data. Although the two sensor networks do not have the interoperability to communicate with one another, they both transfer the monitoring data to the nearby sensor gateway, which makes the system function as a whole.

The WSNs consist of commercially available low cost nodes with small size and low power, which are integrated with sensing and communication capabilities. These elements in our proposed system include nodes with integrated sensors measuring temperature, light, humidity, and combined power switches. The latter power switches are used to control remote power switches to do more things. Historical monitoring data collected from the sensor nodes are stored in the distributed NoSQL data warehouse.

The whole process is divided into three steps. The first step is that sensor nodes collect the sensor data from devices and transfer the monitoring data to the sensor gateway using the communication standards ZigBee and 6LoWPAN. The second step is that the sensor gateway transfers the UDP monitoring data to the server using the WebSocket protocol. We also embed the WebSocket protocol into the chip of sensor gateway. The last step is that the WebSocket compatible server store the historical monitoring data in the distributed NoSQL databases, and push the monitoring data to the WebSocket client.

In order to actually see what is happening when controlling the switches from a remote terminal, a network camera has been connected to the ethernet. This network camera displays in the browser the live monitoring in the building. There are two popular categories of wireless sensor network.

Type 1: ZigBee Wireless Sensor Network
As depicted in Fig. 6.9, ZigBee Wireless Sensor Network is composed of devices from different vendors. The sensor gateway functions both as a server transferring the monitoring data to the WebSocket server, and a connection point between ZigBee WSN and ethernet. It has the coordinating role in the ZigBee network. The script running on the gateway enables the transfer of monitoring data in the ethernet.

We deploy two batteries-driven wireless sensor nodes with internal sensor chips to measure the temperature and humidity in each room. These nodes are able to form mesh networks and can operate within an indoor range of about 20 m. They function as end-devices in the ZigBee WSN. Another sensor node is equipped with a power socket and has a gateway function in the network.

We deploy a remotely controlled ZigBee device with each switch to power the external circuit, such as desk lamps, computers and air-conditioning. They are also used as meters for measuring the load, voltage, current and power of the attached electrical equipment.

Type 2: 6LoWPAN Wireless Sensor Network
The 6LoWPAN WSN is formed by one WebSocket compatible server, two sensor routers, and five battery driven sensor nodes. The architecture of 6LoWPAN WSN is shown in Fig. 6.9. The WebSocket compatible server is connected to the ethernet

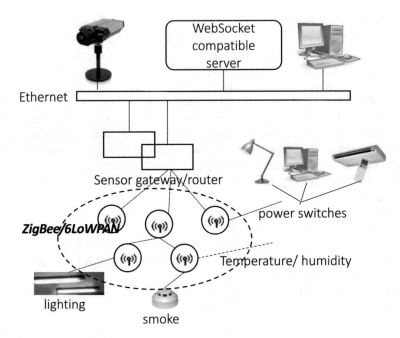

Fig. 6.9 ZigBee/6LoWPAN wireless sensor network

and provides a Web user interface, which stores monitoring sensor data. Through this WebSocket and Web user interface, the user is able to see the real-time monitoring graph and manage to control WSNs of intelligent buildings in the Web browser.

Access to the network from the server goes through the two parallel working sensor routers. These routers manage the routing between the sensor nodes and the IP ethernet. Deploying at least two routers in the same network of sensor nodes will scale the throughput of the network. Each router is able to take over the other router's tasks in cases of a non-operational router, which increases the redundancy and reliability of the network. To utilize this range of usage of the 6LoWPAN sensors, they are connected to a device functioning as a relay controlling the circuits in the building. From the Web interface in the browser, the device was remotely controlled with the circuit switches.

6.4.2.1 WebSocket-Based Real-Time Monitoring

In this section, we discuss some components of the monitoring architecture.

Component 1: Web User Interface
Besides the monitoring based on WebSocket, the proposed system uses HTML5 Canvas API to represent data from WSNs in electronic engineering easily and efficiently. In the proposed HTML5-based monitoring system, sensor gateway can

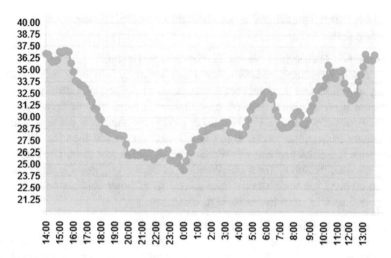

Fig. 6.10 Temperature chart reading from ZigBee sensor nodes

be interacted with and controlled through WebSocket API and their status and functions can be monitored on Canvas. We can perform the deployment of sensor nodes and the configuration management of WSNs in electronic engineering.

WebSocket-based Web clients are JavaScript applications that run in a Web browser, which communicate with the Web and WebSocket server. Canvas API provides scripts with a resolution-dependent bitmap canvas, which can be used for rendering graphs, or other visual images on the fly. We employ open-source Chart. js to represent the data gathered from sensors. Chart.js is an HTML5 JavaScript Charts library, which supports over 20 different types of charts. Using the HTML5 Canvas tag, Chart.js creates these charts in the Web browser, meaning quicker pages and less Web server load. Figure 6.10 shows the Web user interface of temperature movements collected from one sensor of the ZigBee WSN. This Web user interface is implemented with HTML5 Canvas. This Web graph is listed with the accompanying information of the latest sensor data as well as the time of the sensor readings.

Component 2: WebSocket Server versus WebSocket Client
WebSocket requires its own backend application to communicate with the server side. Therefore, we utilize Socket.IO to develop WebSocket server. Socket.IO, plugins of Node.js, aims to easily make real-time applications possible in every browser and mobile device, blurring the differences between the different transport mechanisms, which is carefree realtime 100 % in JavaScript.

On one hand, we take advantage of socket.IO server API to implement a WebSocket server. While the monitoring gateway has received the sensor data, WebSocket server launches an event to notify all the online browsers. On the other hand, we take advantage of socket.IO client API to intercept the message from

WebSocket server in near real-time. The callback function is used to display the monitoring data.

Component 3: WebSocket Server versus Sensor Gateway

Besides monitoring, our WebSocket-based monitoring system for remote intelligent building in WSNs can send real-time commands to the sensor gateway to control the sensors (turn on lights, turn off air-condition, and change camera's sampling frequency) directly through the WebSocket API. Notifications are sent when an event occurs from sensor nodes to the WebSocket server and then the WebSocket server sends these notifications the WebSocket clients in real time. The monitoring actions are sent promptly from the WebSocket server to gateways and/or sensor nodes to respond the notifications. And sensor monitoring data are collected periodically and sent to store in the NoSQL databases.

Component 4: Storage of Historical Monitoring Sensor Data

Figure 6.11 illustrates the overall design architecture of storage of historical monitoring data including all the main components. Storage architecture consists of three main components which are directly contributing to the NoSQL data warehouse. These three main components are: query processor, put processor and MongoDB database.

Query and put processors are the read/write components of the NoSQL data warehouse. Query processor is used to track the historical monitoring data to generate reports or charts, while put processor is used to store the historical monitoring data. We adopt MongoDB as the storage of NoSQL data warehouse, because MongoDB is an open-source leading document NoSQL database. Some features of MongoDB adapt to the monitoring scene. For example, it provides fast query with full index support.

In order to reduce the redundancy rate and storage space of NoSQL data warehouse, we propose a historical storage model with lifecycle. Compared with current timestamp solution, we use the start and end timestamp instead of the unique timestamp.

We give the definition of the lifecycle. The lifecycle tag is a 2-tuple of start and end element, which implies the lifecycle of a record in the NoSQL data warehouse.

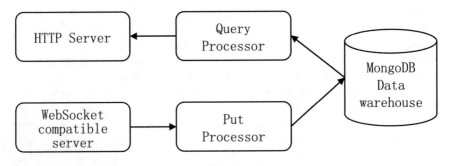

Fig. 6.11 Storage and query processing architecture

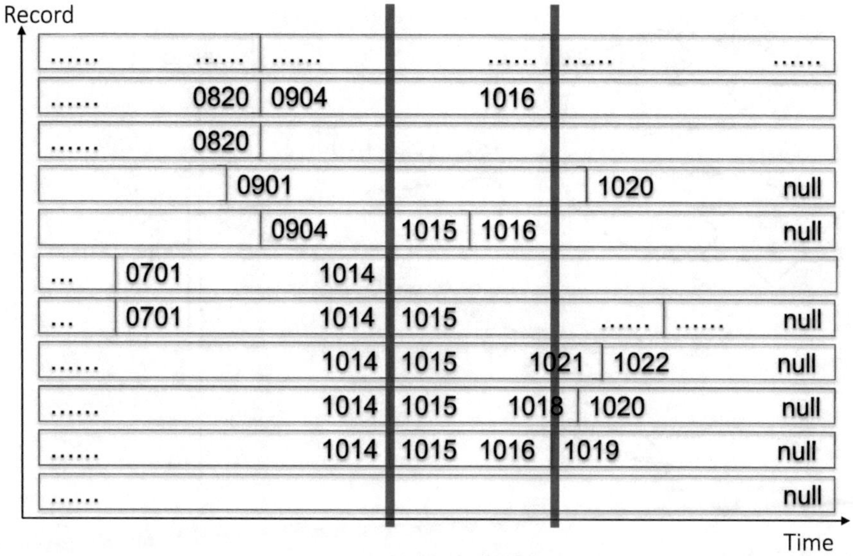

Record

......	......	......		......	......	......
......	0820	0904		1016		
......	0820					
	0901				1020	null
	0904	1015	1016			null
...	0701	1014				
...	0701	1014	1015	......	......	null
......	1014	1015	1021	1022		null
......	1014	1015	1018	1020		null
......	1014	1015	1016	1019		null
......						null

Time

Fig. 6.12 Method to get the historical snapshot over a period of time

The lifecycle is denoted as (start, end). The first element is the start timestamp, and the second element indicates the end timestamp.

We define the historical storage model with lifecycle in the data warehouse as a list of attributes that comprise a key, a set of regular attributes and a lifecycle field. This definition is the metamodel of the historical storage model with lifecycle. That is to say that the record in practice is the instance of this model.

There are two characteristics of this model. First, each record has one and only one lifecycle. Second, each lifecycle belongs to a set of records. In the scenario of the collection of the historical monitoring sensor data, we utilize this storage model to compress the real-time monitoring data.

Figure 6.12 shows the method to get the historical snapshot over a period of time. The storage models with lifecycle that are penetrated by the two black lines formulate the full snapshot of the historical data between the date 1015 and 1017. The snapshot for some time is achieved in this way.

6.5   Evaluation

The experiments were performed on an Intel(R) Core(TM) CPU I5-2300 3.0 GHz server with 8 GB RAM, a 100M Realtek 8139 NIC in 100M LAN. We have made two experiments to illustrate the advantages of our monitoring approach and system. The systems were configured with a Windows Server 2012 x64.

Our first experiment is comparing HTTP polling, Socket in ActiveX, FlashSocket, and our WebSocket monitoring approach. Our experiment is based on

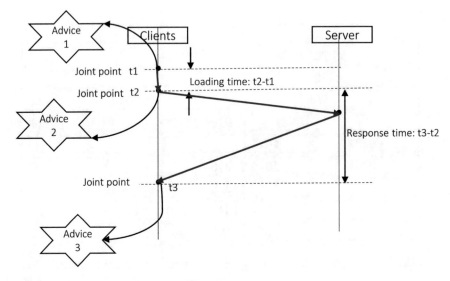

Fig. 6.13 AOP interceptors to calculate response time

aspect-oriented programming (AOP) code that intercepts the methods of the client at the well defined join points to collect test data and measure the loading time and response time at important instants of time.

In order to evaluate the performance of different approaches, we assumed that the startup and shutdown time of AOP interceptors is small enough to ignore. We just calculate the loading time and response time shown in Fig. 6.13. The aspect code measures the timestamp by advice 1, 2 and 3. The loading time is evaluated by the time difference t2–t1, and the response time is evaluated by the time difference t3–t2.

We illustrate four sets of experiments to illustrate the superiority of the developed system.

6.5.1   Fast Loading

We conducted 10 tests to measure the loading time. Since the loading time is calculated by t2–t1, and is independent of the server, we just calculates the average loading time as a reference. Table 6.1 shows the average loading time of different monitoring solutions. Since Socket in ActiveX and FlashSocket depend on the plugins, they have the higher loading time. Due to the similarity to the HTTP protocol, HTTP Polling solution has the lowest loading time. The loading time of our WebSocket solution is the median.

Table 6.1 Comparison of loading time

Solution	Loading time (ms)
HTTP polling	53
Socket in ActiveX	852
FlashSocket	796
WebSocket	121

6.5.2   Low Latency

We conducted a total of two groups of tests for one monitoring process with different solutions. The IP of WebSocket server in the first group is in China to simulate high speed network, while the IP of WebSocket server in the second group is outside China just to simulate low speed network. We have made 5 test use cases to calculate the response time of different solutions. The sampling time is 05:00, 09:00, 13:00, 17:00 and 21:00. Figure 6.14 shows the results of this experiment.

When the network transport is not very satisfying, the latency of FlashSocket and WebSocket solution is similar. When the network transport goes smoothly, the latency of WebSocket solution is lower than FlashSocket solution. Since FlashSocket is embedded in the form of ActiveX or plugins in the browser, the performance of FlashSocket cannot reach their full potential, especially in the low speed network. Although the response time of WebSocket solution is close to FlashSocket solution, it should be noted that the FlashSocket solution needs to download an additional Flash object file before a connection is established. That process will cost extra time. The experimental evidence suggests that the response time of WebSocket shows a little below FlashSocket solution. For the real-time, all the tests show that the HTTP polling and Socket in ActiveX solutions cost significantly longer response time than FlashSocket and WebSocket solution.

Table 6.2 shows the standard deviation of the response time. The standard deviation of the developed WebScoket solution is the lowest.

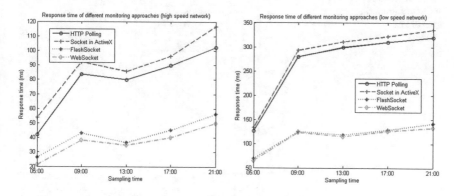

Fig. 6.14 Response time of different monitoring approaches

Table 6.2 Standard
deviation of loading time

Solution	High speed network	Low speed network
HTTP Polling	22.5740	80.4718
Socket in ActiveX	22.8496	83.1940
FlashSocket	11.083	27.9408
WebSocket	10.4430	27.4649

6.5.3   High Concurrency

We conducted the third experiment to test the availability of different monitoring
approaches in different concurrency values. We use multi threads to simulate
concurrent access. Figure 6.15 shows the average response time of different con-
currencies. In the high-concurrency situation, the response time of HTTP Polling
and Socket in ActiveX is so high that the whole system does not work. As the
concurrency counts increase, the response time widens between WebSocket and
FlashSocket solution.

6.5.4   Low Consumption

We conducted the fourth experiment to test the consumption of different monitoring
approaches in different concurrency values. Since all the monitoring approaches are
based on the browser, the CPU utilization of client is one performance criterion. We
have made this experiment to test the average CPU utilization of WebSocket client
accompanied by the last experiment. Figure 6.16 the average CPU utilization in
different concurrency values. The average CPU utilization in WebSocket solution is
almost stable. That is to say that the users do not feel the existence of extra resource

Fig. 6.15 Concurrent test of
different monitoring
approaches

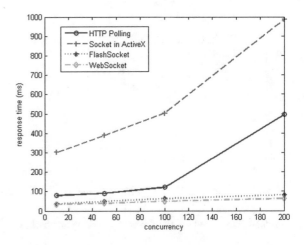

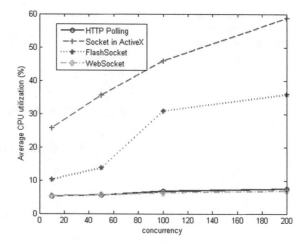

Fig. 6.16 Average CPU utilization of different monitoring approaches

consumption of WebSocket by contrast with the traditional HTTP request and response. FlashSocket and Socket in ActiveX consume a lot of resources because of the embedding plugins techniques, especially in high-concurrency situations.

6.6   Discussions

This section introduces the historical storage model of this architecture for further discussion. We initialize the data warehouse as empty. After that, the collected historical data are stored in the data warehouse. We observed the collected experimental data of timestamp and lifecycle solutions.

6.6.1   Redundancy Rate

First, we conducted the redundancy experiment of timestamp and lifecycle solutions to illustrate the superiority of our approach. Figure 6.12 shows the redundancy rate in 180 days. As depicted in Fig. 6.17, the redundancy rate of the timestamp solution is down with the unchanged probability of the lifecycle solution. Our lifecycle solution is advantageous in the long run.

6.6.2   Storage Space

Next, we analyzed the storage space of different solutions. We assumed that the data size increases by average 5000 documents from the sensor gateway every day. We observed the storage space of the corresponding data warehouse every day.

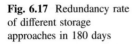

Fig. 6.17 Redundancy rate
of different storage
approaches in 180 days

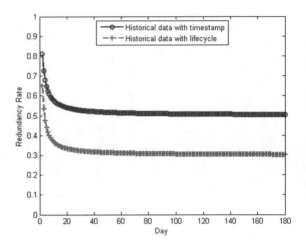

Fig. 6.18 Storage spaces in
180 days

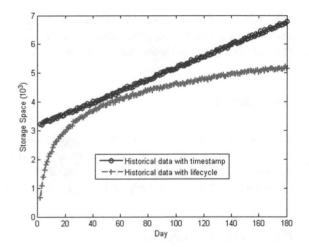

Figure 6.18 shows the variation of different solutions. Scales of the data size of the
corresponding data warehouse increase linearly. The fastest increasing solution is
timestamp solution, since all the records need to be stored regardless of whether the
documents are changed or not. However, our lifecycle solution takes remarkable
superiority in the scenario of big data.

6.6.3   Query Time

In the process of the storage space experiment, we measure the query time of the
historical data in the data warehouse. In order to make the comparison of different
solutions, we select the historical data on the first month. We select 8 points to

Fig. 6.19 Query time of different storage approaches

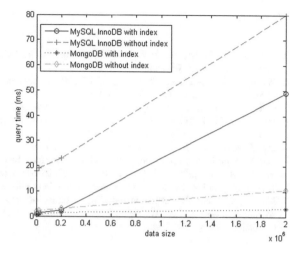

record the time of the same query. Figure 6.19 shows the query time of different solutions. The query time of the lifecycle solution is lower than the timestamp solution all the time. We can conclude that our solution is feasible in practice.

6.7   Conclusions

In this chapter, a WebSocket-based real-time monitoring architecture for remote intelligent buildings is proposed. By utilizing the WebSocket, Canvas, Chart and NoSQL, the Web-based monitoring architecture can be implemented to easily control and monitor the sensor activity in real time. This demonstration illustrated that it is possible to remotely control the electrical appliances from this Web user interface of smart terminals. The open architecture of the concept allows for easy and continuous updates and unlimited expandability. This experimental work presented illustrates that a combination of available WSN and Ethernet can be employed to monitor and measure real time data such as temperature, light, humidity and power consumption. Besides, the experimental results show the efficiency of query processing, especially in a world of big data. The capabilities offered by the type of wireless sensor architecture presented in this chapter are vast. They provide the managers and owners of buildings feedback on the energy consumption of buildings to support improved building control and inhabitant behavioral change. Improvements in the architecture sensors could also be integrated into the type of WSN discussed in this chapter to supply more detailed information to building occupants.

References

1. Yick, J., Mukherjee, B., & Ghosal, D. (2008). Wireless sensor network survey. *Computer Networks, 52*(12), 2292–2330.
2. Wong, J. K. W., Li, H., & Wang, S. W. (2005). Intelligent building research: A review. *Automation in Construction, 14*(1), 143–159.
3. Cattell, R. (2011). Scalable SQL and NoSQL data stores. *ACM SIGMOD Record, 39*(4), 12–27.
4. Juels, A. (2006). RFID security and privacy: A research survey. *IEEE Journal on Selected Areas in Communications, 24*(2), 381–394.
5. Ma, K., & Sun, R. (2013). Introducing websocket-based real-time monitoring system for remote intelligent buildings. International Journal of Distributed Sensor Networks.
6. Ma, K., & Zhang, W. (2014). Introducing browser-based high-frequency cloud monitoring system using websocket proxy. International Journal of Grid and Utility Computing, *6*(1), 21–29.
7. Ma, K., & Yang, B. (2015) Introducing extreme data storage middleware of schema-free document stores using MapReduce. International Journal of Ad Hoc and Ubiquitous Computing, *20*(4), 274–284.

Part IV
Literature Management

Chapter 7
Intelligent Web Data Management of Literature Validation

7.1 Introduction

7.1.1 Background

As the amount of researchers of a scholarly institution increase at an astonishing rate, it becomes both more time-consuming for the researchers to enter the scientific literature (i.e. research chapter results). In the meantime, it becomes more egregiously daunting for scientific departments to audit the authenticity of each literature manually.

7.1.2 Challenges and Contributions

A solution of reliable and automated literature validation architecture would be of great benefit. However, there are many barriers to such similar solutions. The challenge is how to enter the bibliography of its scientific literature conveniently, and how to ensure the entered literature is authentically attached to its owner. To address this limitation, this chapter describes the literature validation architecture using Digital Object Identifier (DOI) [1] content negotiation proxy which we developed.

An important finding of this chapter is that it requires significant development work to combine open CrossCite [2] DOI content service and DOI resolvers of registry agencies. As more than 95 % of DOI is owned or managed by CrossRef [3], DataCite [4], ISTIC [5] and mEDRA [6] DOI registry agencies, it is a common universal approach for the scientific research chapter result with DOI.

© Springer International Publishing Switzerland 2016
K. Ma et al., *Intelligent Web Data Management: Software Architectures and Emerging Technologies*, Studies in Computational Intelligence 643,
DOI 10.1007/978-3-319-30192-1_7

7.2   Related Work and Emerging Techniques

7.2.1   Literature Bibliography Acquisition

In this section, we outline six strategies for literature validation: manual entry bibliography, direct import bibliography, webpage crawl of chapter database, non-scanned document extraction, document ID lookup, and Google Scholar and Microsoft Academic Search API [7].

Method 1: Manual entry bibliography
In the manual approach, bibliography is entered by users. Although that is time-consuming, it is probably the simplest way to enter bibliography in practice.

Method 2: Direct import bibliography
This approach means importing from the formatted citation file (such as EndNote citation file, citation file of research information systems, and BibTeX) directly. The citation file is always saved or downloaded from the online bibliographic database, the university catalog or the CD-ROM. Moreover, Google Scholar now provides the means of exporting individual citations into the formatted textual file. As the typical representative of reference management tools, EndNote provides some personalized filters to import the reference data. Although the entire process is free from the anticipation of people, this approach needs the formatted textual file provided by the third medium.

Method 3: Webpage crawl of chapter database
The webpage detail of the literature database, containing the bibliographic meta-data, is always composed of the back-end structured data and the front-end semi-structured webpage template. Therefore, the metadata extraction of the crawled webpage is another approach to bibliography acquisition. Several research groups have focused on the problem of extracting structured data from the tidy XHTML documents [8]. Most of the research is in the context of semi-automatic and automatic construction of extraction wrappers of the specific webpage template. This approach is with relatively high success rate, but still need manual intervention. The bibliography extractor plug-ins of most current reference management tools (such as Endnote Web [9], RefWorks [10], Zotero [11] and Mendeley [12]) need upgrade or rectifications to maintain such availability in all circumstances. That means that the plug-in is dependent on the webpage template of the chapter database.

Method 4: Non-scanned document extraction
In non-scanned document extraction approach, the bibliographic metadata is extracted from the Office word, latex source document and the PDF document conforming to the template. Since the template is actually not consistent, this approach needs manual intervention with the higher failure rate. For instance, CrosssRef provides an open-source set of tools and libraries for identifying and extracting the significant regions of a scholarly journal article or conference

proceeding PDF. It performs structural analysis to determine column bounds, headers, footers, sections, titles and so on. It can analyze and categorize the sections into reference and the non-reference sections, and also can split reference sections into individual references. Another example is that the bibliographic metadata can be extracted from the PDF file mixed with the pre-generated XMP file. The big advantage of embedding bibliographic metadata in the PDF is that the content and metadata are never separated. However, the disadvantage of this approach is that not all the publishers and other content producers embed metadata into their PDFs. An improved scheme of this approach is to identify the DOI of none-scanned document, and then the bibliography is extracted from the DOI content negotiation.

Method 5: Document identifier lookup

Document identifier lookup approach is looking up document details from document ID system, such as digital object identifier (DOI), PubMed unique identifier (PMID) and ArXiv. In the most popular document ID, DOI becomes an ISO standard in May, 2012. ISO 26324:2012 specifies the syntax, description and resolution functional components of the digital object identifier system. Take for example the largest DOI registration agency CrossRef, there are more than 66,681,470 DOI available by April 14, 2014 [13]. That means that DOIs are heavily used in the publishing space to identify electronic documents uniquely, especially scholarly journal and conference articles. Although PMID and ArXiv are also digital document ID, they are confined to specific fields. For instance, PMID refers to about the life sciences and biomedical topics, and ArXiv refers to the fields of mathematics, physics, astronomy, computer science, quantitative biology and statistics.

Method 6: Google Scholar and Microsoft Academic search API

Google Scholar and Microsoft Academic search engine provide APIs to complete the missing data and build the authentic ranking of chapters. It is an additional solution to acquire the bibliography. The disadvantage of this approach is the heterogeneity of API and lack of the automatic validation of the bibliography. In addition, the search results may not be the latest and overall indeed.

7.2.2   DOI Content Negotiation and Resolver

On the base of the above methods, DOI, the most common digital document identifier, provides more than a permanent and indirect link to content. DOI registration agencies, such as CrossRef and DataCite, collect bibliographic metadata (author, affiliation, title, source, volume and issue number, starting page number, and abstract et al.) about the works they link to.

One resolution of this approach is DOI content negotiation [14], which is used to retrieve different representations of a work. Content negotiation is a mechanism defined in the HTTP specification that makes it possible to serve different versions of a document (or more generally, a resource representation) at the same URI, so

Table 7.1 Metadata content types of DOI content negotiation

Format	Content type
RDF XML	application/rdf+xml
CrossRef Unixref XML	application/vnd.crossref.unixref +xml
DataCite XML	application/vnd.datacite.datacite +xml
BibTeX	application/x-BibTeX
...	...

that user agents can specify which version fit their capabilities the best. It allows a user to request a particular representation of a web resource. In this method the user will not access this service directly. Instead, the user makes a DOI resolution via dx.doi.org using an HTTP client which allows to specify HTTP Accept header. Content negotiation for DOI names is a collaborative effort of CrossRef and DataCite. We discuss metadata content types of DOI content negotiation in detail in Table 7.1. Using content negotiation it is possible to make a request that favors content types specific to a particular registration agency. The advantage of this resolution is that content negotiation lets you process a DOI's metadata without knowledge of its origin or specifics of the registration agency. In 2011, the DOI registry agencies CrossRef and DataCide announced to support unified content negotiation. However, this approach is not at the more general end of the scale. Some other registration agencies, such as the Institute of Scientific and Technical Information of China (ISTIC) and the Multilingual European DOI Registration Agency (mEDRA), do not support content negotiated DOIs. Another resolution of this approach is DOI resolver. It is the interface for a specific registration agency to obtain the bibliographic metadata or webpage from DOI. For example, the DOI resolver at dx.doi.org will normally redirect a user to the resource location of a DOI. However, the interface of DOI resolver is heterogeneous, and each of the registration agencies has its own specification. The challenge is how to integrate the resolver together and design a DOI resolver for the new registry agency.

To address this limitation, we design a literature validation architecture using DOI content negotiation proxy to combine with these above two resolutions. In addition, the user interface supporting both PC and smart devices are designed to allow the researcher to add the bibliography by simply using the DOI, which would then fill in the other bits of bibliographic metadata needed for display. As this metadata is achieved from the DOI resolver rather than entered manually, this mechanism ensures the authenticity of the literature.

7.2.3　Bibliographic Model

There are plenty of contenders though, each based on differing models for how this data should be encapsulated. For example, RDF provides the bibliographic ontology with a way to mix together different descriptive vocabularies in a consistent way. Vocabularies can be created by distinct communities and groups as appropriate and mixed together as required, without needing any centralized agreement on how terms from different vocabularies can be written down in XML. The main characteristic difference is in how markedly hierarchical or flat the model structure is. A model that has emerged from the library world is Functional Requirements for Bibliographic Records (FRBR). FRBR is built around the notion of books—what a book is, taking into account things like editions and so on. Some experts proposed FaBiO ontology to encapsulate the information in a FRBR-like manner.

Besides RDF, there are some other bibliographic models, such as Citeproc JavaScript Object Notation (JSON), BibTeX and Research Information Systems (RIS). Citeproc JSON is a JavaScript bibliographic model rendered by the citeproc-js citation processor; BibTeX stands for a tool and a file format which are used to describe and process lists of references, mostly in conjunction with LaTeX documents; RIS is a standardized tag format developed by Research Information Systems, incorporated to enable citation programs to exchange data. All the bibliographic models are dependent on the specific implementation, and they can be converted into their own formats. To solve this problem, this study proposed a platform-independent model named BibModel to describe the bibliography at utmost. All the specific models can be generated from transformation.

7.3   Requirements

We have the following requirements while designing a literature validation architecture.

Literature validation: We want to validate the literature researchers enter using a bibliography acquisition approach.

Integration with the third-party systems: We want to become the service providers of literature validation to integrate with the third-party systems using friendly interface.

7.4   Architecture

In this section, the architecture and its implementation details of bibliography acquisition approach using DOI content negotiation proxies are discussed in detail. The architecture is shown in Fig. 7.1.

As for DOI resolution, there is no uniform norms and technical specifications to follow. We propose an integration solution with the heterogeneous interface of DOI

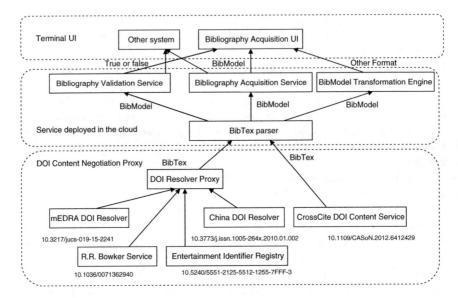

Fig. 7.1 Literature validation architecture

registry agency to provide a uniform service. In our solution, DOI resolver proxy provide an interactive and independent interface no matter what the implementation of the interface is. For the existing interface of DOI registry agency, we just integrate it with our framework. While for the DOI registry agency without DOI interactive interface, we have made some development work to implement DOI resolution first. The contribution of DOI resolver proxy in our framework is attempting to solve the heterogeneity problem of current DOI resolvers, which can also provide second-development interface for anyone who is interested in it. DOI content service is another supplement of DOI resolution, which is based on HTTP content negotiation. However, this interface style is not supported by other DOI registry agencies except CrossCite (CrossRef and DataCite). In our framework, we implement CrossCite DOI content negotiation service.

7.4.1   Bibliography Acquisition Architecture

The bibliography acquisition architecture is composed of three points of view in order to reduce the complexity: terminal UI, service deployed in the cloud, and DOI content negotiation proxy. The terminal user interface of bibliography acquisition is used to enter the DOI or the title of chapter, supporting both traditional PC and smart devices; service deployed in the cloud provides not only the interface of bibliography acquisition and validation, but also BibTeX parser from BibTeX to

BibModel; DOI content negotiation proxy is a proxy who collects bibliography from diffident DOI content negotiations or DOI resolvers.

7.4.2 DOI Content Negotiation Proxy

The heterogeneity of the service interface of DOI registry agency determines the existence of DOI content negotiation proxy. The flow chart of DOI content negotiation proxy is shown in Fig. 7.2. As most of DOI is managed by CrossRef and DataCite (hereinafter abbreviated to CrossCite), the acquisition request is first redirected to the CrossCite DOI content service. And then, the untreated request is sent to the component DOI prefix judgment for the next process. This component can judge the affiliation by the DOI prefix and redirect the request to the right resolver. Finally, the metadata is obtained.

The DOI content negotiation proxy we designed is to provide the bibliography from the DOI, no matter what the registry agency is. As the new input DOI, the proxy is first to determine whether this DOI is in CrossCite or not. If yes, it is requested to the CrossCite DOI content negotiation; if not, it is requested to other DOI resolver proxy.

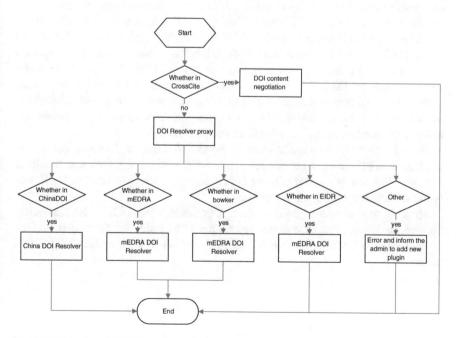

Fig. 7.2 Flow chart of DOI content negotiation proxy

7.5   Evaluation

This section evaluates the literature validation architecture from the perspective of functionality.

7.5.1   DOI Content Service

The DOI Content Service framework introduces Accept header defined in the HTTP specification to specify certain media types, which are acceptable for the response. The most commonly format is application/rdf+xml, application/x-BibTeX and text/x-bibliography. We do some extra significant development work based on the current CrossCite interface.

Currently, the two largest DOI Registration Agencies CrossRef and DataCite support a shared, common metadata format to provide metadata services to DOI names. The service will make an HTTP content negotiation request to the global DOI resolver specifying which format of the metadata should be returned in the HTTP accept header. The global DOI resolver will notice Accept header that this is not a regular DOI resolution request. It will turn to CrossRef or DataCite accordingly for the relevant metadata instead of redirecting to a landing page. For example, a client that wishes to receive BibTeX if it is available, but which can also handle RDF XML if BibTeX is unavailable, would make a request with an accept header listing both "application/x-BibTeX" and "application/rdf+xml". This request favors BibTeX but will accept RDF XML if BibTeX is unavailable. The Sequence Diagram of DOI content negotiation proxy is shown in Fig. 7.3, and the response code is shown in Table 7.2. Individual metadata service of registry agencies may utilize additional response codes, but they will always use the response codes above in the event of the case described. As for the code 204, we rectify a new redirection to the DOI resolver proxy introduced in the next section.

In our solution, we take advantage of net package of JAVA programs to implement HTTP content negotiation. The JAVA method negotiation is shown as follows. With the parameter url and HTTP accept header, we can get DOI resolution of different bibliographic records.

Here is an example of DOI content negotiation. The DOI is 10.1155/2014/583686, and the result of content negotiation is as follows. The output of DOI content negotiation may be BibTeX directly, which is as the input of BibTeX parser.

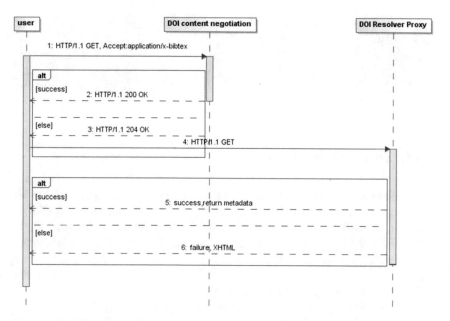

Fig. 7.3 Sequence diagram of DOI content negotiation proxy

Table 7.2 Response codes of DOI content negotiation

Code	Meaning
200	The request was OK
204	The request was OK but there was no metadata available
404	The DOI requested doesn't exist
406	Can't serve any requested content type

curl-LH "Accept: application/rdf+xml;q=0.5, application/x-bibtex;q=1.0"
http://dx.doi.org/10.1155/2014/583686

Result:
@article{Ma_2014,
title={Multiple Wide Tables with Vertical Scalability in Multitenant Sensor
Cloud Systems},
volume={2014},
ISSN={1550-1477},
url={http://dx.doi.org/10.1155/2014/583686},
DOI={10.1155/2014/583686},
journal={International Journal of Distributed Sensor Networks}, publisher=
{Hindawi Publishing Corporation},

```
author={Ma, Kun and Yang, Bo},
year={2014},
pages={1–10}}
```

7.5.2   DOI Resolver Proxy

CrossCite DOI Content Negotiation provides the most mapping from DOI to bibliography. With the collaboration of the largest DOI registry agencies CrossRef and DataCite, CrossCite owns more than 95 % of the DOI. As for the bibliography not in CrossCite, the DOI resolver is designed to extract the BibTeX from the results of DOI resolvers. The specific DOI resolver is just like the flexible plug-in mounting to the proxy.

Here are two examples of DOI resolvers.

Take mEDRA (the multilingual European Registration Agency of DOI) for example. mEDRA provides the Web service to view the metadata of bibliography. The WSDL of this service is as follows. The method viewMetadata is obtaining the bibliographic metadata by the means of DOI, with the input doi and output contentID.

```
<message name="viewMetadataRequest">
  <part name="doi" type="xsd:string"/>
</message>
<message name="viewMetadataResponse">
  <part name="contentID" type="xsd:base64Binary"/>
</message>
<portType name="MedraWSType">
<operation name="viewMetadata" >
<input name="inputViewMetadata" message="tns:viewMetadataRequest"/>
<output name="outputViewMetadata" message="tns:viewMetadataResponse"/>
</operation>
</portType>
```

In our solution, we take advantage of Apache CXF framework to implement Web service interface. We use WSDL2Java of CXF to generate the client MedraWSService and service interface MedraWSType. MedraWSService and MedraWSType are shown as follows.

```
@WebServiceClient(name = "MedraWSService",
        wsdlLocation = "medra.wsdl",
        targetNamespace = "http://www.medra.org")
public class MedraWSService extends Service {
  ...
  @WebEndpoint(name = "medraWS")
  public MedraWSType getMedraWS() {
   return super.getPort(MedraWS, MedraWSType.class);
  }
}
@WebService(targetNamespace = "http://www.medra.org", name = "Medra
WSType")
@SOAPBinding(style = SOAPBinding.Style.RPC)
public interface MedraWSType {
    @WebResult(name = "contentID", targetNamespace = "http://www.medra.
org", partName = "contentID")
    @WebMethod(action = "viewMetadata")
public byte[] viewMetadata(
      @WebParam(partName = "doi", name = "doi")
      java.lang.String doi
    );
}
```

We can use the method viewMetadata to implement DOI resolver as follows.

```
MedraWSService service = new MedraWSService();
MedraWSType client = service.getMedraWS();
byte [] result = client.viewMetadata("10.1392/roma081203");
```

Take another DOI registry agency ISTIC for instance. It can facilitate the query and positioning of the bibliography conforming to OpenURL specification. If you find the metadata of DOI 10.3773/j.issn.1005-264x.2010.01.002, the OpenURL is as follows. The bibliographic metadata could be parsed from the XML results. In order to improve readability of the result, some Chinese word has been translated into English.

```
curl–L http://www.chinadoi.cn/openurl.do?pid=wf:wf
&id=doi:10.3773/j.issn.1005-264x.2010.01.002&noredirect=

Result:
<chinesedoi_result version="1.0.0">
<query_result>
<body>
<query>
<journal>
<journal_metadata>
<full_title>CHINESE JOURNAL OF PLANT ECOLOGY</full_title>
<issn media_type="print">1005-264X</issn>
<cn media_type="print">11-3397/Q</cn>
</journal_metadata>
<journal_issue><journal_volume><volume>34</volume>
</journal_volume><issue>01</issue></journal_issue>
<journal_article>
<titles><titlet>Ecological stoichiometry: Searching for unifying principles
from individuals to ecosystems</titlet></titles>
<contributor>Jinsheng He</contributor>
<doi_data><doi>10.3773/j.issn.1005-264x.2010.01.002
</doi><resource></doi_data>
<pages><first_page>2</first_page></pages>
</journal_article>
</journal>
</query>
</body>
</query_result>
</chinesedoi_result>
```

In our solution, we take advantage of net package of JAVA programs to implement OpenURL resolution. With the parameter url, we can get bibliographic records conforming to OpenURL specification.

7.5.3   DOI Presentation

DOI system was designed to provide a form of persistent identification, in which each DOI name unequivocally and permanently identifies the object to which it is associated. The common way of presentation is DOI string, just like 10.1155/2014/583686. Another presentation is a persistent link, just like http://dx.doi.org/10.1155/2014/583686. The last presentation is DOI QR Code. QR Codes are a form of

Fig. 7.4 QR code of DOI
persistent link

barcode that can be easily scanned and used by mobile phones, web-cams, etc.
Inspired by Google's recent promotion of QR Codes, we experiment with encoding a
DOI and a bit of metadata into one of the critters. The QR Code will include the title
of the document pointed to by the CrossRef DOI as well as the "http://dx.doi.org"
URL of that will allow you to link to that item. Minimally, this DOI QR code is used
to help the researchers improve the input efficiency of mart devices. An example of
QR Code 10.1155/2014/583686 is shown in Fig. 7.4. In our framework, QRCode is
stored as binary data.

7.5.4   BibTeX Parser

BibTeX parser can parse a BibTeX-file and render it as part of BibModel. The
advantage of BibModel rather the concrete format such as RDF and BibTeX is used
to simplify the bibliographic acquisition service interface. In our solution, we take
advantage of JBibTeX API to implement parsing BibTeX to render BibModels.

7.5.5   Bibliography Validation Service

Our proposed framework can integrate with the current management system of the
scientific research results. The entire process is the following: first of all, the user
information such as the user name can be obtained from the context session on the
management system of scientific research results. Second, the researcher enters the
DOI of its bibliography, and another metadata can be obtained from our framework.
At last, the integrated system compares author's name in the context and author's
name obtained from DOI content negotiation proxy. They indicate the bibliography
entered is authentic attached to its owner if they are consistent. In bibliography
validation service of our framework, validateBib is to validate the authenticity
through the author name extracted from the metadata. With the maturity and

improvement of DOI technology in the future, the information is also validated by the affiliation of the metadata too. This design of validation service ensures that the researcher cannot input the bibliography not attached to its owner.

7.5.6 BibModel Transformation Engine

Only the BibModel output data is provided with the service deployed in the cloud other formats can be generated from the BibModel transformation engine with the

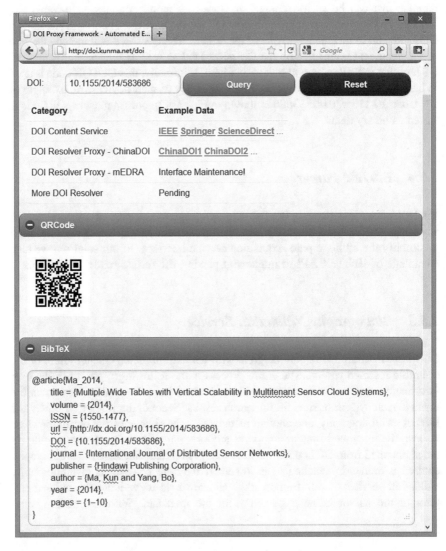

Fig. 7.5 User interface of the prototype system using this approach supports PC and smart mobiles

approach. In order to generate the understandable bibliographic record, the implementation detail is designing the concrete template.

In our solution, we proposed the template execution engine to implement BibModel transformation. Since the template statement is composed of text, interpolation, tag and comment, the key of implementing template execution engine is replacing interpolation expression and executing tag statements. As for the interpolation expression delimited by ${ and }, the expression is just replaced after model transformation. As for the tag statements, the predefined directives and user-defined directives are invoked to replace the statement with the executed result.

7.5.7 Terminal UI

The user interface of the prototype system using this approach supports PC and smart devices, as shown in Fig. 7.5. The DOI can also be entered manually or captured using the camera of smart devices, such as mobile phones and tablet computers. We use the Adobe PhoneGap to implement the above features, which creates mobile applications for iOS, Android, Blackberry, Windows Phone, Palm WebOS, Bada and Symbian using the web code.

7.6 Discussions

This Section discusses bibliographic models and the relational transformation method as further extensions of the literature validation architecture.

This heterogeneity of current bibliographic models (such as Citeproc JSON, BibTeX and RIS) has brought challenges to unify the representation of bibliography. This study proposes an abstract bibliography model named BibModel to describe the bibliographic record in the best possible way. BibModel is enough of an abstract model to render an understandable bibliographic record to support a number of metadata content types, such as RDF XML, RDF Turtle, Citeproc JSON and BibTeX. BibModel is independent of the implementation details, generating the concrete bibliographic record after model transformation.

Model transformation is the bridge between our BibModel and concrete bibliographic records. Model transformation is the process of converting one model to another model of the same system. The input to the transformation is the marked BibModel and the mapping. The result is the specific bibliographic models.

7.6.1 Bibliographic Model—BibModel

To be able to define the bibliography acquisition, a mathematical representation of the bibliographic model is necessary. In our study, we proposed BibModel using first-order predicate logic. The key/value ordered pair of bibliographic attribute is a function BibAttribute: $K \rightarrow V$ and can be rewritten as BibAttribute(k)=v where $k \in K$, $v \in V$. Here, k represents the key of the bibliographic attribute, while v means the value of the bibliographic attribute. The bibliographic model is defined as the mathematical relation, denoted as BibModel: <T, {a|a∈BibAttribute}>, where T is the bibliographic type, generally ranged from article, book, proceeding and the thesis. The relation BibModel satisfies the following constraints. The bibliographic type determines the numbers of bibliographic attributes. The commonly type of bibliographic model is listed in Table 7.3. The predicate dom means the domain of function BibAttribute. In order to describe the constraints, the projection operator is introduced to present the i_th projection of n-ary sequence, denoted as i_th(a1,a2, ..., an)=ai. In order to present the model, consider that the predicate META(o, c)=true means that o is the instance of c. In other words, the data type of o is c. Compared with BibModel, Table 7.4 shows examples of concrete bibliographic record.

Table 7.3 The commonly used type of bibliographic model

Type	Meaning
Article	An article from a journal or magazine
Book	A book with an explicit publisher
Inproceedings	An article in conference proceedings
Mastersthesis	A Master's thesis
Phdthesis	A Ph.D. thesis
Techreport	A report published by a school or other institution
Misc	For use when nothing else fits. It is suitable for electronic resources

Table 7.4 Examples of concrete bibliographic records

BibTeX	JSON	RIS
@inbook{Ma_2010, title={test}, booktitle={CSCWD 2010}, publisher={IEEE}, author={Ma, Kun}, year={2010}, pages={71–76}}	{"title":"test", "container-title":"CSCWD 2010", "publisher":"IEEE", "author":[{"family":"Ma", "given":"Kun"}], "page":"71–76", "type": "chapter-conference"}	TY—CONF T2—CSCWD 2010 AU—Ma, Kun TI—test SP—171 EP—76 PB—IEEE PY—2010

∀META(bib, BibModel)=true ∧ 1_th(bib)=Article ∃**dom**(2_th(bib))={author, title, journal, year, volume, number, pages, month}

∀META(bib, BibModel)=true ∧ 1_th(bib)=Book ∃**dom**(2_th(bib))={author, title, publisher, year, volume, series, address, edition, month}

∀META(bib, BibModel)=true ∧ 1_th(bib)=Inproceedings ∃**dom**(2_th(bib))= {author, title, booktitle, year, editor, volume, series, pages, address, month, organization, publisher}

∀META(bib, BibModel)=true ∧ 1_th(bib)=Mastersthesis ∃**dom**(2_th(bib))= {author, title, school, year, type, address, month}

∀META(bib, BibModel)=true ∧ 1_th(bib)=Phdthesis ∃**dom**(2_th(bib))={author, title, school, year, type, address, month}

∀META(bib, BibModel)=true ∧ 1_th(bib)=Techreport ∃**dom**(2_th(bib))= {author, title, institution, year, type, number, address, month}

∀META(bib, BibModel)=true ∧ 1_th(bib)=Misc ∃**dom**(2_th(bib))={author, title, note, howpublished, year}

7.6.2   Transformation from BibModel to Bibliographic Records

BibModel is the intermediate transferred on the Web. With the model transformation, it can render an understandable bibliographic record. In this section, a template-based bibliographic model transformation is presented. The input is BibModels and templates, and the output is specific bibliographic records.

A template is a series of template statements. The set of transformation rules from BibModel to the target bibliographic models is denoted as Rule = $\sum$templatei. The model transformation rule of the evolution of textual template is based on all the template statements.

A template statement is defined as 4-tuple: TemplateStatement:=<Text, Interpolation, Tag, Comment>.

The text is static and it will keep constantly after model transformation.

The interpolation will be replaced with a calculated value in the output, which is the dynamic content of templates. The format of interpolations is ${expression}, where interpolations are delimited by ${ and }. The result of the expression must be a string, number or date value. This is because only numbers and dates will be converted to string by the interpolation automatically, other types of values (such as Booleans and sequences) must be converted to string manually somehow, or an error will stop the template processing.

The tag introduces some evolution mechanism to satisfy the requirements for the specific application field, such as macro, iteration, condition and function statements. The tag is used to call directives. There are two kinds of tags: start-tag and end-tag. Generally speaking, the format of start-tag is <#directivename parameters>, while the format of end-tag is </#directivename>. The format of the parameters depends on the directivename. This is similar to HTML or XML syntax, except that the tag name starts with #. In fact, there are two types of directives: predefined directives and user-defined directives. For user-defined directives, you use @ instead of #, for example <@mydirective parameters>...</@mydirective>. Further difference is that if the directive has no nested content, you must use a tag like <@mydirective parameters/>, similarly as in XHTML (e.g. <img ... />).

There are four kinds of directives as shown in Fig. 7.6: macro, condition, list and function. Macros work by simple textual search-and-replace at the token. Its syntax is: <#macro macroname param1, param2,..., paramN> macro block </#macro>, where macroname is the name of macro variable, and param1, param2,..., etc. are the names of the local variables store the parameter values; conditional directives are used to conditionally skip a section of the template based on the predicate of the Boolean expressions. Its syntax is: <#if BooleanExpression> template statement 1<#else> template statement 2</#if>; list directives are used to process a section of template for each variable contained within a sequence. For each such iteration, the loop variable will contain the current sub variable. Its syntax is: <#list sequence as item>template statement</#list>; a function is to create a method variable. This directive works in the same way as the macro directive, except that return directive must have a parameter that specifies the return value of the method, and that attempts to write to the output will be ignored. If the </#function> is reached (i.e. there was no return returnValue), then the return value of the method is an undefined variable. Its syntax is: <#function functionname param1, param2,..., paramN>...<#return returnValue>...</#function>. functionname is the name of method variable, and param1, param2,..., etc. are the names of the local variables

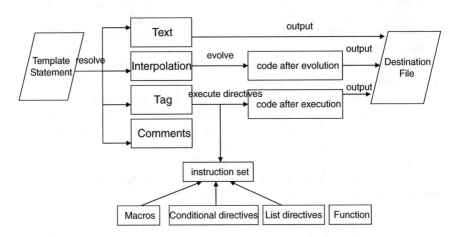

Fig. 7.6 Template directives

Table 7.5 The template from BibModel to concrete bibliography records

BibModel to BibTeX	BibModel to JSON	BibModel to RIS
@${1_th(bib)} {@${1_th(bib)}, <#list dom(2_th(bib)) as K> ${K}="${BibAttribute (K)}", </#list> }	{<#list dom(2_th(bib)) as K> "${K}"= "${BibAttribute(K)}", </#list> }	TY-${1_th(bib)} <#listdom(2_th(bib)("author"))as author>AU-${author}</#list> TI-${2_th(bib)}.title PB-${2_th(bib)}.publisher SP-${2_th(bib)}.page.prefix EP-${2_th(bib)}.page.postfix PY-${2_th(bib)}.year UR-http://dx.doi.org/${2_th(bib)}.doi

store the parameter values, and returnValue is the expression that calculates the value of the method call.

The comment is similar to HTML tag comment, but it is delimited by <#– and –>. Comments will be ignored and not be written to the output.

An example of the model transformation templates from BibModels to BibTeX, JSON and RIS are shown in Table 7.5. Consider bib is an instance of BibModel, META(bib, BibModel)=true. For the generation of other formatted bibliographies such as RDF and JSON, the only difference is the content of template.

We make some explanation of BibModel to BibTeX template in the first column of Table 7.4 as a typical example. We use the statements of template language in Sect. 3.2. ${1_th(bib)} is the interpolation statement, which is replaced with the first projection of instance bib after model transformation. The next few rows in the first column of Table 7.4 go through the bibliographic attributes of instance bib, generating the key/value pairs of BibTeX. The other JSON and RIS templates are analogous to this example.

The contribution of BibModel provides a general and platform independent bibliographic model in an abstract way. With transformation from BibModel to bibliographic records, BibModel can generate concrete bibliographic records. The heterogeneity problem of bibliography can be resolved to a certain extent.

7.7 Conclusions

As the amount of the scientific research chapter result increases at an astonishing rate, a scheme for biography acquisition would be of urgent need. We proposed an approach using DOI content negotiation proxy. The architecture and its implementation are discussed in detail. In order to simplify the service interface and support many kinds of biographic formats, we presented the independent BibModel and its template-based model transformation engine to support the rich biography records. An important finding is that we designed the framework combining open CrossCite DOI content service and DOI resolvers of registry agencies. However, there is significant development work to be done while integrating these services.

As more than 95 % of DOI is owned or managed by CrossRef, DataCite, ISTIC and mEDTA DOI registry agencies, this is a common universal approach for the scientific research article retrieval result using DOI.

References

1. Paskin, N. (2008). Digital object identifier (DOI) system. *Encyclopedia of Library and Information Sciences, 3*, 1586–1592.
2. CrossCite. https://www.datacite.org/node/63
3. CrossRef. http://www.crossref.org
4. DataCite. https://www.datacite.org
5. ISTIC. http://www.chinadoi.cn
6. mEDRA. https://www.medra.org/
7. Fenn, J. (2006). Managing citations and your bibliography with bibtex. The PracTEX Journal (4).
8. Wong, T. L., & Lam, W. (2010). Learning to adapt web information extraction knowledge and discovering new attributes via a bayesian approach. *IEEE Transactions on Knowledge and Data Engineering, 22*(4), 523–536.
9. Gomis, M., Gall, C., & Brahmi, F. A. (2008). Web-based citation management compared to EndNote: Options for medical sciences. *Medical Reference Services Quarterly, 27*(3), 260–271.
10. Marsalis, S., & Kelly., J. (2008). Building a RefWorks database of faculty publications as a liaison and collection development tool.
11. Zotero. https://www.zotero.org.
12. Mendeley. http://www.mendeley.com.
13. CrossRef indicators. http://crossref.org/01company/crossref_indicators.html.
14. Ma, K., & Yang, B. (2014). A simple scheme for bibliography acquisition using DOI content negotiation proxy. *The Electronic Library, 32*(6), 806–824.

Chapter 8
Intelligent Web Data Management of Literature Sharing

8.1 Introduction

8.1.1 Background

Access to scientific literature information is a very important and time-consuming daily work for scientific researchers. Although this growth has allowed researchers to quickly access more scientific information, it has also made it more difficult for them to find literatures relevant to their interests, and keep academic exchange with their counterparts. Therefore, modern researchers need new unified and general approaches and tools for sharing literature available to them.

Historically, there are some legacy literature sharing approaches [1, 2].

- The first way is that researchers find articles only by following citations in other articles that they are interested in. This is an effective practice, but it limits researchers to specific citation communities, and it is biased towards heavily cited chapters.
- The second way is to find articles by online Scientific publisher Web sites (example: IEEE Xplore [3], SpringerLink [4] and Elsevier ScienceDirect [5]), and secondary literature databases (i.e. Engineering Index Compendex [6], Science Citation Index [7], and Elsevier Scopus [8]). A complementary method of finding articles is based on the keyword search, which is a powerful approach.
- The last approach is to share the literature in the third-party system. In this way, researchers have to enter or import the literature to the system and share these with other researchers. Recently, Web sites like CiteULike [9] and Mendeley [10] allow researchers to create their own reference libraries for the articles they are interested in and share the literature with other counterparts. This has opened the door to use recommendation methods as a third way to help researchers find interesting articles. However, this approach needs the researchers to acquaint with the third-party system. The challenging is the integration of various heterogeneous systems.

© Springer International Publishing Switzerland 2016
K. Ma et al., *Intelligent Web Data Management: Software Architectures and Emerging Technologies*, Studies in Computational Intelligence 643,
DOI 10.1007/978-3-319-30192-1_8

8.1.2 Challenges and Contributions

All the current approaches have their limitations. First, current literature sharing approaches lack of the support of academic exchange. They are on the basis of the literature rather the researchers. Second, the extra work using the third-party system makes the sharing more complicated.

There are some challenges of intelligent Web data management of literature sharing. The first challenge of the literature sharing is how to push the sharing to the researchers rather find the sharing by themselves. Another challenge is how to do the academic exchange with the other researchers with the same interests. Therefore, we urgently need a new novel tool to create the opportunities of online archives to inform researchers about the literature that they might not be aware of but interested in.

Recently, cloud computing has the potential to transform a large part of traditional application to make the software even more attractive as a service. On one hand, we need some cloud services rather than other third-party systems in the context of literature sharing. That is to say that cloud architecture has the flexibility, enabling users to utilize the service without installing any software. On the other hand, all the data are stored in the cloud. Therefore, we design a cloud architecture to implement the unified and general literature sharing services.

In this chapter, we attempt to change the traditional passive search literature sharing into actively pushing the literature when the researchers browse the Web page. Cloud services are assisted to complete the literature sharing. Furthermore, we develop a sidebar tool in the cloud, integrated with the browser. This tool is triggered by the bookmarklet to share the literature. Each researcher can see the sidebar by clicking the bookmarklet, on which the relevant literature and communication interfaces are. The content on the sidebar is obtained by the Digital Object Identifier (DOI), which provides more than a permanent and indirect link to content. We use the cloud DOI resolver developed to extract the metadata of bibliography. Besides, we present the relevant documents that cite this literature on the sidebar automatically. This process is triggered by the user with a very simple and intuitive action: the selection with the mouse on a piece of text of the current Web page, and the clicking on the bookmarklet.

Compared with current sharing approaches, we use bookmarklet to control the display of the sidebar. The contributions of this chapter are several folds. First, a bookmarklet is a platform independent tool that enables users to visually share the literature presented in a Web document in an unobtrusive way. This approach is unified since it is suitable for all the heterogeneous publishers. Second, this approach works across browser platforms without installing additional plugins and software. Researchers can share the literature to do the academic exchange in the publisher page with the help of cloud services. Third, we use emerging Web 2.0 techniques (i.e. WebSocket) to implement real-time online academic exchange. Finally, we use NoSQL to store comments and reviews of the literature to provide the lowest latency display.

8.2   Related Architecture and Emerging Techniques

8.2.1   Emerging Web Technologies

Web 2.0 is basically the trend in the use of World Wide Web (WWW) technology and web design that aims to enhance the creativity, information sharing, and collaboration among users. These concepts have led to the development and evolution of web-based communities and hosted services. It is playing a vital role in the literature sharing. Therefore, we use bookmarklet, WebSocket and NoSQL to benefit the literature sharing.

A bookmarklet is a bookmark stored in a Web browser that contains JavaScript commands to extend the browser's functionality. For instance, a bookmarklet might allow the user to select the DOI on the Web page of the publisher, click the bookmarklet, and then be presented with a sidebar of the literature sharing with the academic exchange interface. Whether bookmarklet utilities are stored as bookmarks or hyperlinks, they are designed to add one-click functionality to a browser or Web page. When clicked, a bookmarklet performs some function, one of a wide variety such as a search query or data extraction.

WebSocket is a Web technology in the HTML5 specification providing full-duplex communications channels over a single TCP connection. In this way, the WebSocket protocol enables the interaction between a browser and a Web server, which is difficult to implement using the traditional HTTP protocol. There are some popular WebSocket applications. In this chapter, we use WebSocket as the academic exchange protocol to make researchers communicate with each other in real time. In contrast to the HTTP request-response model, the WebSocket interface enables continuous bi-directional communication between browsers and servers. The significance of this bidirectional communication mechanism is that a browser can now be thought of as a real-time terminal for back-end applications. From this perspective, this is made possible by providing a standardized way for the server to send content to the browser without being solicited by the client, and allowing for messages to be passed back and forth while keeping the connection open.

NoSQL enhances the overall performance of applications that exhibit a high read to write ratio. For the purpose of the presentation of the comments and reviews, we use NoSQL to store the related data to provide the effective query interface.

8.2.2   Literature Sharing

Currently, there are at least five methods for sharing literature. The first approach is the literature publisher system. Each publisher has their own publishing systems, in which the sharing, comments, and communication areas are displayed. For instance, Springer provides the related content area for all the literature published. The second approach is the third-party document management integrated system. After logging in this system, researchers are able to share the literatures entered or imported by themselves. The domestic China's Sciencechapter Online and New Science are two typical examples of online document management systems. The third approach is an academic forum and bulletin board system, which implements the literature sharing whenever you post or replies the threads. China's Emuch Research Forum is an example of this category. The fourth approach is the public sharing in the forms of microblog, blog and personal space. This approach shares the URL of the publisher Webpage of the literature. For instance, the Springer China publisher provides the sharing link to Sina microblog. The last approach is the push message component. This component can push the URL of the literature to the specific instant messaging client. For example, Tencent provides "share to QQ" component interface on the Web. Researchers can click the "share" button to push the link of the literature to the friend or the group of QQ.

The above classical literature sharing approaches have some limitations. The literature publisher system is open to its end users, and close to other users. The users in different publisher systems cannot communicate with each other. The third-party document management integrated systems and academic forums only serve its users. All the three approaches need to reach a consensus on the literature sharing platforms. Moreover, URL-based approaches lack of the unified standard and specification.

In this chapter, we proposed a unified literature sharing approach triggered by the bookmarking. Bookmarklet can be seen as an early form of the literature sharing. For example, Zotero is an open-source competitor that is built around a widget that works with many popular Web browsers. Mendeley is a new service that is gaining acceptance not only as a citation manager but also as a means to annotate and share electronically retrieved research in PDF format. Although the approach we present is similar to the above methods, the difference is that our approach is integrated with the publisher Web site in the form of Cloud services transparently. That is to say that researchers are able to share the literature as they browse the publisher Web page.

8.3   Requirements

We have the following requirements while designing a literature sharing architecture.

Researcher-oriented rather than literature-oriented: In the process of literature sharing, we fully mobilize the subjective initiatives of researchers. We expect the researchers using this system become friends or at least know each other in case of browsing the publisher page of the literature.

Easy to Handle and interoperability: We expect to simplify the process of literature sharing. Furthermore, we enable the end users of this system to communicate with each other in real time. This approach is unified, independent of the different publisher systems.

Low latency of communication: Since the researchers are busy, we expect that someone will respond to the target sharer with low latency, just like a chatting room.

Transparency of the sharing system: We do not want the end users in fact to feel the existence of the sharing system. Therefore, we store the sharing data of the literature in the Cloud transparently.

8.4 Architecture

In this section, we discuss the architecture of our proposed bookmarklet-triggered unified literature sharing in the Cloud.

8.4.1 Hierarchical Model of Bookmarklet

First, we introduce the hierarchical model of bookmarklet. As depicted in Fig. 8.1, the bookmarklet interacts with the remote objects instead of interacting with the native JavaScript objects. After the user clicks a bookmarklet, it triggers the remote objects to generate the executed JavaScript running in the context of the local page dynamically. Due to security issues, we use JSONP [11] to implement the request from a server in a different domain, something prohibited by typical web browsers because of the same origin policy.

8.4.2 System Architecture

Second, we introduce the architecture of literature sharing. As depicted in Fig. 8.2, the architecture is composed of five points of view in order to reduce the complexity: browser, bookmarklet, sidebar client, cloud process engine, and cloud storage engine. The browser is the client of Web-based literature sharing. The bookmarklet is a bookmark stored in a Web browser that contains JavaScript commands to trigger the process of the literature sharing. Sidebar client displays the related information of the literature. Besides, the academic exchange WebSocket

Fig. 8.1 Hierarchical model of bookmarklet

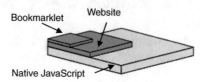

Fig. 8.2 Literature sharing architecture

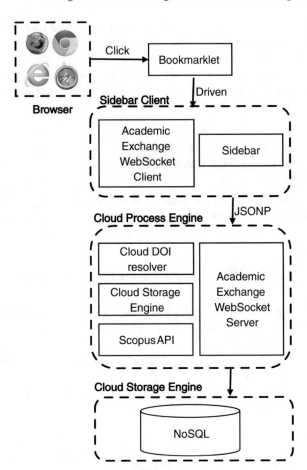

client that is used to implement the communication interface lies on the sidebar client. Cloud process engine is the process engine in the cloud. We use cloud DOI resolver, cloud storage engine and Scopus API to provide the cloud service for the sidebar client. Furthermore, the academic exchange WebSocket server that is used to implement the message forward lies in the cloud. All the related data are stored in the cloud NoSQL database.

8.4.3   Literature Sharing Process

The tentative sharing process for the bookmarklet-triggered unified literature sharing consists of the following steps, as shown in Fig. 8.3.

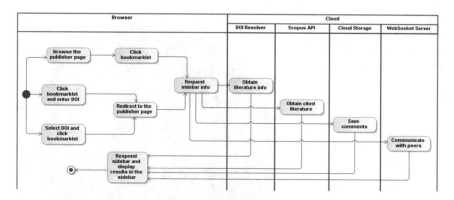

Fig. 8.3 Bookmarklet-triggered unified literature sharing process

We provide three simple ways to trigger the programs of bookmarklet. The first way is clicking the bookmarklet in the case of browsing the publisher page. The second way is clicking the bookmarklet and entering DOI of the literature. The last way is selecting the DOI and clicking the bookmarklet. The last two ways will redirect the current Web page to the publisher.

A standard bookmark consists of two parts: a URL and a bookmark name. Instead of a standard URL, a bookmarklet uses JavaScript to become a type of mini-program. These brief programs can do a variety of actions to display the sidebar with the information obtained from the Cloud. Bookmarklets can work in most popular browsers. Researchers control the display of the sidebar by invoking the scripts previously stored in their bookmarks. The bookmarklet is available from a setting menu on their home page. It contains a token that uniquely identifies the user.

Next, we request the Cloud process engine for the literature information and its cited document based on the current DOI in the Cloud. The literature information is obtained from the Cloud DOI resolver. The cited literature is obtained using Scopus API. We save the comments and reviews of the current literature in the Cloud NoSQL storage engine. We provide the communication area for the academic exchange using a WebSocket protocol with low latency.

8.5   Evaluation

The experiments were performed on an Intel(R) Core(TM) CPU I5-2300 3.0 GHz server with 8 GB RAM, a 100 M Realtek 8139 NIC in 100 M LAN. We have made several experiments to illustrate the advantages of our sharing approach and service. The services were deployed within a Windows Server 2012 x64.

Fig. 8.4 User satisfaction of
our service

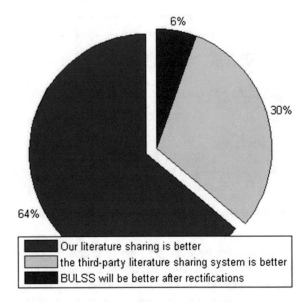

8.5.1 *Bookmarklet Versus Third-Party Platform*

We now turn to quantitative evaluation in terms of user satisfaction. We made a
survey to the researchers of University of Jinan, Shandong, China. We asked the
researchers to rank the two systems: one is a famous third-party literature sharing
system, another is our system. Due to licensing restrictions, we cannot disclose the
identity of the third-party literature sharing system. Figure 8.4 shows the feedback
of user satisfaction. We have received 424 valid feedbacks. 79.2 % of researchers
thought that our system is more friendly, 17.7 % of researchers felt that the
third-party literature sharing system is better, and 3 % of researchers voted that our
system may be better after some rectifications. Generally, the users who dislike our
system are used to the traditional literature sharing system. We believe they would
attempt our system after the promotion. The approach of our literature sharing has
some advantages. First, this system saves time to simplify the literature sharing.
Second, our system can push the related literatures to the researchers. Finally, we
provide a set of API to be integrated with other third-party systems.

8.5.2 *WebSocket Versus FlashSocket*

We have made the first experiment to compare the academic exchange approach
using FlashSocket and WebSocket respectively, which is based on aspect-oriented
programming (AOP) code that intercepts the methods of client and server at the
well defined join points to measure the timestamps at important instants of time.

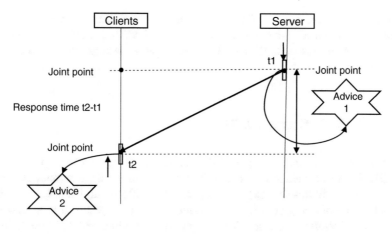

Fig. 8.5 AOP interceptors to calculate response time

In order to evaluate the performance of different approaches, we are assuming that the startup and shutdown time of AOP interceptors is small enough to ignore. The aspect code calculates the timestamp by advice 1 and 2. Moreover, the response time is evaluated by the time difference t2–t1, which is shown in Fig. 8.5.

Figure 8.6 shows the results of this experiment. We ran a total of 2 groups of tests for one interactive process with different solutions. The IP of WebSocket server in the first group is in China to simulate high speed network, while the IP of WebSocket server in the second group is outside China just to simulate low speed network. We have made 5 test use cases to calculate the response time of different solutions. When the network transport is not very satisfying, the latency of FlashSocket and WebSocket solution is similar. When the network transport goes smoothly, the latency of WebSocket solution is lower than FlashSocket solution. Since FlashSocket is embedded in the form of ActiveX or other plugins in the

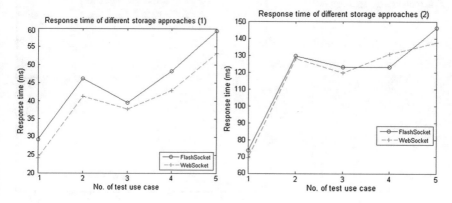

Fig. 8.6 Response time of different approaches

browser, the performance of FlashSocket cannot reach their full potential. It will become the bottleneck when the network is very slow. The experimental evidence suggests that the response time of WebSocket shows a little below FlashSocket solution.

8.5.3   NoSQL Versus RDBMS

We save the comments, reviews and academic exchange records in the Cloud NoSQL. The distributed system determines that we have a large amount of records in the database. In this section, we compare the RDBMS with NoSQL to illustrate the superiority of our approach. We have made the first experiment to measure the query time of RDBMS and our NoSQL solution. We adopt 64 bits MySQL 5.6.14 GA as the RDBMS, and MongoDB 2.4 as the NoSQL database. We calculate the query "select the academic exchange data in the last 12th hours" to observe the response time. We initialize about 2000, 20,000, 200,000 and 2,000,000 items in the MySQL as well as MongoDB respectively. Although the structure of the data is different, the substantial data are the same. In order to optimize the structure of MySQL and MongoDB, we have added the index of the search field. We make the above test use case of query to calculate the execution time. Figure 8.7 shows the query time. Since the current database has a cache for the query result, we just calculate the query time for the first time. It indicates that the query performance of MongoDB outweighs MySQL, especially in big data.

Next, we have made the second experiment to measure the throughput capacity of MySQL and MongoDB. In this case, we perform different queries to observe the throughput. The experiments confirmed our hypothesis. As depicted in Fig. 8.8, the MongoDB storage solution has the maximum throughput all the time.

Fig. 8.7 Query time of different storage approaches

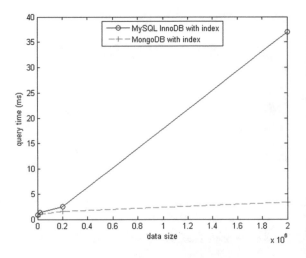

Fig. 8.8 QPS of different
storage approaches

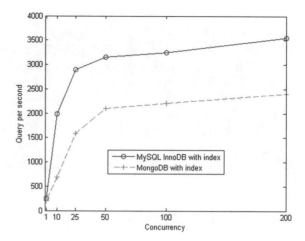

8.6   Discussions

8.6.1   *Bookmarklet*

We will take the second way of the bookmarklet trigger for example, which passes
the parameter DOI to the Cloud service. The content of the bookmarklet is shown as
follows. This bookmarklet inserts the dynamic JavaScript to the current page to
invoke the display of the sidebar.

```
(function(d){if(!!d){d.toggle();return;};
var src =
'http://[Cloud]/bookmarklet.js?doi='+escape(prompt('input%20your%
20doi','DEFAULT DOI'));
var e=document.createElement('script');
e.setAttribute('src',src);
document.getElementsByTagName('head')[0].appendChild(e);
})()
```

In the code segment (see the following) of bookmark.js, it generates a new
translucent DIV layer called sidebar with the close button in the current Web page.
The contents of the sidebar are obtained from the Cloud process engine. They are
introduced in the next few sections. The code segment of bookmark.js is shown as
follows. First, we initialize the modal window modal and its content iframe.
Second, we display this layer on the top of the page in modal way. The content of
the sidebar is inside the iframe tag.

```
(function() {
...
var $iframe = $('<iframe></iframe>').prop({'src': 'http://[Cloud sidebar]/
?doi='+doi});
var $modal = $('<div></div>').append($iframe);
$('body').prepend($modal);
$modal.animate({'opacity':1}, 400);
}());
```

8.6.2   Cloud DOI Resolver

First, we design the interface CloudDOIResolver. We use the service interface of
DOI registration agency (implementation class CloudDOIResolver) to implement
the cloud service for the sidebar clients. As shown in Fig. 8.9, this component is
called Cloud DOI Resolver.

Next, as most of DOI is managed by CrossRef and DataCite (hereinafter
abbreviated to CrossCite), we design the first implementation class CrossCiteDOI
ContentNegotiation using CrossCite DOI content negotiation service. It introduces
accept header defined in the HTTP specification to specify certain media types,
which are acceptable for the response. We do some extra significant development
work with the output header format "application/x-BibTeX". The JAVA method
resolve is shown as follows. With the parameter doi, we can get DOI resolution of
different bibliographic records.

```
public String resolve(String doi)throws Exception{
URLConnection conn = new URL("http://dx.doi.org/"+doi).openConnection
();
conn.setRequestProperty("accept","application/x-bibtex");
BufferedReader in = new BufferedReader(new InputStreamReader(conn.
getInputStream(),"UTF-8"));
while ((inputLine = in.readLine()) !=null)sb.append(inputLine).append("\n");
in.close();return sb.toString();
}
```

Fig. 8.9 Class diagram of
cloud DOI resolver

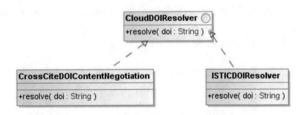

Finally, we design another implementation class ISTICDOIResolver using Institute of Scientific and Technical Information of China (ISTIC) DOI service API. It can facilitate the query interface of the ChinaDOI system. The bibliographic metadata could be parsed and extracted from the results of the Web page. The JAVA method resolve is shown as follows. With the parameter doi, we can get bibliographic records.

```
public String resolve(String doi)throws Exception {
Document    doc=Jsoup.connect("http://www.chinadoi.cn/chinadoi-manage/
manage/doiBatchParse.action").data("doiStr",    doi).userAgent("Mozilla/4.0
(compatible; MSIE 6.1; Windows XP)").cookie("JSESSIONID", cookies.get
("JSESSIONID")).cookie("wangfang",    cookies.get("wangfang")).timeout
(30000).post();
Element element=doc.getElementsByClass("col_ff0").get(0);
return extractBibTex(element);
}
```

The Cloud DOI resolver we design is to provide the literature information by the DOI, no matter what the DOI registration agency is. As for the new DOI registration agency, the only thing to do is adding a new implementation class of interface CloudDOIResolver.

8.6.3   Cloud Storage Engine

In the comment and review areas of the sidebar, we use NoSQL as the data storage engine in the Cloud. The comment refers to the comment on the page of the publisher. Researchers can select the text on the page of the publisher and comment on the sidebar. The review refers to personal remarks. All the filled-in comments and reviews are sent to the Cloud process engine, and then saved in the cloud storage NoSQL. Researchers can see the latest comments and reviews on the sidebar.

We design the query and put processor. Query processor plays a major role in the database abstraction. This is a novel approach for displaying the comments and reviews on the sidebar, because it can solve the query bottleneck issues in the case of big data.

8.6.4   Scopus API

We use Scopus Application Programmer Interface (API) to get the cited literature of the current one via DOI. The core object of this API is searchObj. We use

JavaScript method getciteinfo to display the cited results on the sidebar. After invoking, the results are displayed on the sidebar with animation.

```
function getciteinfo(doi){
$('#citebutton').addClass('ui-disabled');
$("#sciverse").append('Loading...').hide().slideDown("slow");
var varSearchObj=new searchObj();
varSearchObj.setSearch(doi);
sciverse.setApiKey("API KEY");
sciverse.setCallback(function(){$('#citebutton').removeClass
('ui-disabled');});
sciverse.search(varSearchObj);
}
```

8.6.5   Academic Exchange WebSocket Server

In our solution, we take advantage of the SockJS server API to implement the academic exchange WebSocket server. While it has received the request that comes from the sidebar of the researchers, it will launch an event to notify all the online researchers. The following is the abbreviated code segment of the WebSocket server.

```
var http=require('http');
var sockjs=require('sockjs');
var echo=sockjs.createServer();
echo.on('connection', function(conn) {
    conn.on('data', function(message) {
    conn.write(message);});
conn.on('close', function() {});
});
```

8.6.6   Sidebar

As shown in Fig. 8.10, the sidebar is triggered by the bookmarklet. It is divided into five parts: literature information, cited literature, comment sharing, review sharing, and real-time academic exchange. Literature information area displays the detail information of current literature of the publisher. Cited literature area displays the

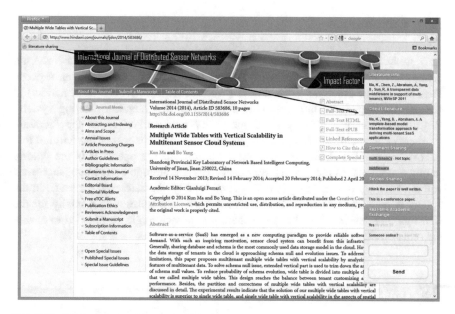

Fig. 8.10 Sidebar triggered by the bookmarklet

related article that cites this literature. Comment sharing area displays the comments of the current Web page. Compared with comment sharing, review sharing area displays the remarks of different researchers on the literature. Real-time academic exchange area displays the communication interface in real time, which functions as the chatting room of researchers with low latency.

8.6.7 Academic Exchange WebSocket Client

In our solution, we take advantage of the SockJS client API to intercept the message from academic exchange WebSocket server to display the interactive interface on the sidebar in near real-time. Compared with the primitive WebSocket, SockJS is a browser JavaScript library that provides a WebSocket-like object, which provides a coherent, cross-browser Javascript API to support a low latency, full duplex, cross-domain communication channel between the browser and the Web server. The callback function we design is shown as follows.

```
var sock=new SockJS('http://[Cloud WebSocket Server IP]');
sock.onmessage=function(e) {
   $("#message").append(e.data).slideDown("slow");
};
```

8.7   Conclusions

This study aimed at introducing bookmarklet-triggered unified literature sharing architecture in the cloud. The main features are mostly in the following aspects. First, we use the emerging technologies (WebSocket, NoSQL and bookmarklet) to design and implement the researcher-oriented unified literature sharing system. Researchers can use the sidebar triggered by bookmarklet to do the literature sharing and academic exchange in the browser. Second, the researcher will not feel the existence of the cloud system using our approach. The operation of the sharing in our system makes the sharing very easy. The literature is on the publisher itself without entering or importing to our system. Finally, the academic exchange in the system is with low latency and network flow. These features are especially for the smart mobile devices. This literature sharing approach might be pursued to enable the other similar sharing to benefit from this technology.

References

1. Bouyukliev, I., & Georgieva-Trifonova, T. (2013). Development of a personal bibliographic information system. *The Electronic Library, 31*(2), 144–156.
2. Ma, K., & Yang, B. (2014). A simple scheme for bibliography acquisition using DOI content negotiation proxy. *The Electronic Library, 32*(6), 806–824.
3. IEEE Xplore. http://ieeexplore.ieee.org.
4. SpringerLink. http://link.springer.com.
5. Elsevier ScienceDirect. http://www.sciencedirect.com.
6. Engineering Index Compendex. http://www.engineeringvillage.com.
7. Science Citation Index. http://science.thomsonreuters.com.
8. Elsevier Scopus. http://www.scopus.com.
9. CiteULike. www.citeulike.org.
10. Mendeley. http://www.mendeley.com.
11. Ippolito, Bob. "Remote json-jsonp." (2005).

Printed in the United States
By Bookmasters